A Short Guide to Writing about Biology

EIGHTH EDITION

JAN A. PECHENIK
Biology Department
Tufts University

Boston Columbus Indianapolis New York San Francisco Upper Saddle River
Amsterdam Cape Town Dubai London Madrid Milan Munich
Paris Montreal Toronto Delhi Mexico City Sao Paulo Sydney
Hong Kong Seoul Singapore Taipei Tokyo

To my family

Sangria hare: a rain novel, a dear

Senior Sponsoring Editor: Katharine Glynn
Senior Marketing Manager: Sandra McGuire
Senior Supplements Editor: Donna Campion
Production Project Manager: Clara Bartunek
**Project Coordination, Text Design,
and Electronic Page Makeup:** Integra Software Services
Cover Design Manager: Jayne Conte
Cover Designer: Bruce Kenselaar
Cover Image: Image Quest Marine
Printer, Binder: Edwards Brothers
Cover Printer: Lehigh/Phoenix, Hagerstown

For permission to use copyrighted material, grateful acknowledgment is made to the copyright holders which appear on appropriate page within the text.

Library of Congress Cataloging-in-Publication Data
Pechenik, Jan A.
A short guide to writing about biology / Jan A. Pechenik.—8th ed.
p. cm.
ISBN-13: 978-0-205-07507-2 (alk. paper)
ISBN-10: 0-205-07507-X (alk. paper)
1. Biology—Authorship. 2. Report writing. I. Title.
QH304.P43 2013
808'.06657—dc23

2011037788

10 9 8 7 6 5 4 3 2—EB—15 14 13 12

ISBN 10: 0-205-07507-X
ISBN 13: 978-0-205-07507-2

Why Do You Need this New Edition?

If you're wondering why you should buy this new edition of *The Short Guide to Writing about Biology*, here are ten good reasons:

1. Updated information on conducting Web searches and using Open Access Journals for research.
2. Additional instruction on evaluating Web sites offers guidance for one of the most problematic challenges that students face—deciding which online sources are credible.
3. Increased emphasis on issues of plagiarism: what it is and how to avoid doing it.
4. Most chapters now begin with a What Lies Ahead section, preparing you for all that follows in the chapter.
5. Material on writing summaries, critiques, essays, and review papers has been combined into a single chapter.
6. A new overview to Part II now provides a compelling rationale for undertaking these assignments with conviction and motivation.
7. New discussion of using Advanced Search functions in locating sources.
8. New information about specialized note-taking software.
9. More information about the distinction between primary and secondary literature.
10. And finally, the essay question on burger characteristics (Ch. 8) has been updated—at considerable digestive cost!

Contents

Preface

Careful thinking cannot be separated from effective writing. Being a biologist is not just about memorizing facts and terminology, or about mastering an increasing array of computer software and molecular techniques. Biology is a way of thinking about the world. It is about making careful observations, asking specific questions, designing ways to address those questions, manipulating data thoughtfully and thoroughly, interpreting those data and related observations, reevaluating past work, asking new questions, and redefining older ones. It is also about communicating information—accurately, logically, clearly, and concisely. The hard work of thinking about biology is at least as important as the work of doing it. Writing provides a way to examine, to evaluate, to refine, and to share that thinking. Writing is both a product and a process.

Biology instructors are increasingly concerned about their students' writing for two reasons. First, bad writing often reflects fuzzy thinking, so questioning the writing generally guides students toward a clearer understanding of the biology being written about. Second, effective communication is such a key part of the biologist's trade that our students really must learn to do it well. The difficulty, of course, is finding the time to teach both biology and presentation skills when there is barely enough time in the semester to cover the biology. This book allows instructors to guide their students' writing without taking up valuable class time. And as their writing improves, so, too, will the students' understanding of what they are writing about.

Although this book covers every sort of writing assignment that biologists face—both as students and as professionals—it is brief enough to be read along with other, more standard assignments and straightforward enough to be understood without additional instruction. The book is intended especially for undergraduate use in typical lecture and laboratory courses at all levels, but it is also widely used in undergraduate and graduate seminars. Many colleagues tell me they have also found much in the book that was new and helpful in their own writing, and in their teaching.

I have included examples from all fields of biology. However, because the book is intended for use even at introductory levels, I have avoided examples that assume substantial specialized knowledge or terminology. Instructors in advanced courses may wish to amplify basic principles with examples chosen from papers published in their own fields; students will benefit in particular from guided study of good models.

CHANGES MADE FOR THE EIGHTH EDITION

For this edition of the *Not-So-Short-Guide*, I have made the following major changes:

- Each chapter now begins with a brief overview of the main points to be covered in that chapter.
- The material on plagiarism has been consolidated and expanded and is now dealt with in the opening chapter.
- I have updated the Technology Tips when appropriate.
- I have updated the information about conducting Web searches, and now discuss open access research journals (the DOAJ—Directory of Open Access Journals) as a new area to explore for research sources.

- I now discuss specialized note-taking software, and how to narrow literature searches to particular domains and file types using Advanced Search functions in Web browsers such as Google and Scirus.
- I now have more to say about the distinction between the primary and secondary literature.
- I have added an overview to Part II, pulling together several themes that recurred previously in several chapters and eliminating redundancy.
- Chapters 7 and 8 from the previous edition have been combined into a single chapter dealing with the writing of summaries, critiques, essays, and review papers, to emphasize the connections among these assignments and eliminate redundancy: Writing short critiques, for example, is excellent preparation for writing a meaningful review paper.
- In Chapter 9, the section about writing a Results section now makes it clear that maps, drawings, and photographs are also considered figures and therefore need instructive figure captions.
- I have expanded the chapter on answering essay questions, and now relate it more clearly to material concerning the writing of summaries and critiques. Indeed, the chapter has been moved earlier in the book to clarify that connection.
- The chapter on writing essay exams (Chapter 11) now follows this new combined chapter, to show students how the general skills involved in summarizing and synthesizing can be applied to answering exam questions.
- In the concluding chapter, Chapter 12, on writing letters of application, I now emphasize the importance of linking your research interests with the research activities of potential mentors when applying to graduate programs.
- I have also consolidated several of the appendixes, and eliminated others. For example, a single appendix now lists both in-print and Web resources, and a single appendix now shows the corrections to the "sentences in need of revision" from Chapter 6.
- I have launched a companion Web site: http://ase.tufts.edu/biology/faculty/pechenik/ (click on Writing about Biology). Several reviewers have noted that technology tips can quickly become outdated, and that not all students use Word and Excel, to which most of the technology tips apply. Therefore, I plan to set up a Web site for Technology Tips updates from readers; if you wish to submit a Tip for posting, please be sure to indicate at the start the software to which the Tip applies.
- If all goes well, the new Web site will also include links to complete meeting posters, for class discussion. A number of faculty have suggested that I include examples of complete posters in the book, but the small-book, single-color format makes that difficult to accomplish in a helpful way. So here is my solution. If you think you have an outstanding poster design that should be included on the Web site, please send me a copy (PDF) for consideration.
- The new Web site will also include the PDF on Science Journalism from the sixth edition. This chapter was eliminated from the book several editions ago. However, I still believe that having students write for general audiences is an effective teaching tool, forcing students to be clear in both their writing and in their thinking.
- Finally, I also expect the new Web site to include additional material for instructors, in particular suggestions for novel sorts of assignments—with the justification for giving such assignments—along with downloadable templates for use in peer review or for giving students instructor feedback on lab reports and oral presentations.

For this eighth edition, I have retained the narrative style that has made earlier editions so successful with students. We can't expect students to become better writers if we reduce everything to bullets and summaries for them. Students can learn a great deal by writing their own summaries, but little of lasting value by reading or memorizing mine. I have, however, added more boldfacing to this edition, making it easier for students to locate advice of particular importance. Users of the previous edition will notice many smaller improvements in every chapter.

ORGANIZATION

The first 6 chapters cover general issues that apply to all types of writing (and reading) in biology. In Chapter 1, I emphasize the benefits of learning to write well in biology, describe the sorts of writing that professional biologists do, and review some key principles that characterize all sound scientific writing. In Chapter 2, I describe how to locate useful sources using computerized indexing services, online journals, and the Internet. Chapter 3 emphasizes the struggle for understanding that must precede any concern with *how* something is said. In it, I explain how to read the formal scientific literature, including graphs and tables; how to take useful notes; and how to take them in ways that prevent unintentional plagiarism. Chapter 4 talks about the use and interpretation of statistical analyses, while Chapter 5 explains how to cite references and prepare a Literature Cited section. Chapter 6 focuses on the process of revision—for content, organization, clarity, conciseness, grammar, word use, and spelling. It emphasizes the benefits of peer review, and it explains both how to be an effective reviewer of other people's writing and how to interpret criticism. Many readers have found Chapter 6 to be one of the most important chapters in the book. Most students learn very little in preparing the first draft of anything. However, they can learn much—both about biology and about communicating their thoughts—through properly guided revision.

The rest of the book covers all of the specific writing tasks encountered in biology coursework and in professional life: writing summaries, critiques, essays, and review papers (Chapter 7); answering essay questions on exams (Chapter 8); writing laboratory and other research reports (Chapter 9); writing research proposals (Chapter 10); preparing oral and poster presentations (Chapter 11); and writing letters of application for jobs or for graduate school (Chapter 12). I encourage instructors to incorporate short oral presentations into their course design. Writing typically improves when students are first asked to give a short oral presentation on some aspect of what they are planning to write about: Writing, thinking, and speaking are all interconnected. Requiring brief oral presentations is a particularly good way to get students started on larger projects early in the semester.

My discussion of writing summaries and critiques is an especially important part of the book, because most students seem not to have had much practice summarizing information accurately and concisely, and in their own words. An inability to summarize effectively is a serious obstacle to both synthesis and evaluation. Writing summaries is also a particularly effective way for students to self-test their understanding and to prepare for examinations.

The chapter on writing research reports (Chapter 9) emphasizes that the results obtained in a study are often less important than the ability to discuss and interpret those results convincingly in the context of basic biological knowledge, and to demonstrate a clear understanding of the purpose of the study. It emphasizes the variability inherent in biological systems and how that variability is dealt with in presenting, interpreting, and discussing data. This chapter will also be useful to anyone preparing papers for publication.

The checklists found at the ends of most chapters allow students to evaluate their own work and that of their peers. Most of the checklists include page numbers, helping students locate the text on which each item is based. Instructors can easily turn these checklists into grading rubrics, which should be shared with students well before the assignments are due.

"Technology Tips" are scattered throughout the book, helping students take better advantage of the computer technology available to them for finding sources, writing, graphing, and giving oral presentations.

ACKNOWLEDGMENTS

This edition has benefited greatly from the suggestions of many people who took the time to read and comment on the previous edition including Andrea Aspbury, Texas State University; Michael J. Klein, James Madison University; Gwynne Stoner Rife, The University of Findlay; Denise Strickland, University of Virginia; Briana Crotwell Timmerman, University of South Carolina; and Janice Voltzow, University of Scranton.

Previous editions benefited greatly from sage advice given by the following readers: Andrea Aspury, Vickie Backus, Alison Brody, Diane Caporale-Hartleb, Bob Dick, Jean-Marie Kauth, Kirk A Stowe, Alexa Tullis, Barbara Frase, Caitlin Gabor, Blanche Haning, Jean-Marie Kauth, Angie Machniak, R. Brent Thomas, Cathy Tugmon, Christine E. Jungklaus, Scott Kinnes, Chris Maher, Kay McMurry, William R. Morgan, Regina Raboin, Linda L. Tichenor, Cheryl L. Watson, Virginia Anderson, Tina Ayers, Sylvan Barnett, Arthur Buikema, Edward H. Burtt, Robert Chase, Maggie de Cuevas, Robert Curry, John R. Diehl, George F. Edick, Stephen Fuller, Louis Gainey, Jr., Sharon Hanks, Marcia Harrison, Jared Haynes, Joseph Kelty, Scott Kinnes, Anne Kozak, George Labanick, Martin Levin, Sara Lewis, Barbara Liedl, John W. Munford, Colin Orians, Peter Pederson, Laurie Sabol, Carl Schaefer, Christopher Schardl, Stephen van Scoyoc, Barbara Stewart, Marcia Stubbs, and David Takacs.

I am grateful to all of these people for their comments and suggestions and am much cheered by their dedication to the cause. And who could ask for a more attentive reader than Victoria McMillan? "Oh, shame! Where is thy blush?"

It is also a pleasure to thank Regina Raboin of Tufts University, who was a great help in updating the material on conducting Internet and database searches.

Finally, I have learned much about writing and teaching from correspondence and conversation with enthusiastic readers of previous editions, from conversations with faculty in the many workshops that I've led over the past 20+ years, and from working with colleagues from all disciplines in what was once the Writing Across the Curriculum program at Tufts University. I welcome additional comments from readers of the present edition, both instructors and students (**jan.pechenik@tufts.edu**).

JAN A. PECHENIK

I GENERAL ADVICE ABOUT WRITING AND READING BIOLOGY

1

INTRODUCTION AND GENERAL RULES

What appears as a thoroughly systematic piece of scientific work is actually the final product: a cleanly washed offspring that tells us very little about the chaotic mess that fermented in the mental womb of its creator.

AUNER TREININ

What Lies Ahead? In This Chapter, You Will Learn

- The importance and benefits of learning to write well in biology courses
- Twenty-seven rules that characterize all good scientific writing—learn them, and follow them
- The perils of plagiarism and how to avoid them
- How to get the most from your computer in writing, data storage, data analysis, and data presentation

The logical development of ideas and the clear, precise, and succinct communication of those ideas through writing are among the most difficult, but most important, skills that can be mastered in college. Effective writing is also one of the most difficult skills to teach. This is especially true in biology classes, which often require much writing to be done but allow little time to focus on doing it well. The chief messages of this book are that developing your writing skills is worth every bit of effort it takes, and that biology is a splendid field in which to pursue this goal.

WHAT DO BIOLOGISTS WRITE ABOUT, AND WHY?

Most biologists write lectures; grant proposals; research papers; literature reviews; oral and poster presentations for meetings; letters of recommendation; committee reports; and even critiques of research papers, research proposals, and books written by other biologists. The writing that biologists do is similar in many respects to the writing of essays, literature reviews, term papers, and laboratory reports that you are asked to do while enrolled in a typical biology course. Basically, we must all prepare arguments.

Like a good term paper, research report, oral presentation, or thesis, a lecture is an argument; it presents information in an orderly manner, and it seeks to convince the audience that this information fits sensibly into some much larger story. Putting together a string of 3 or 4 lectures on any particular topic is the equivalent of preparing one 20- to 30-page term paper weekly.

In addition to preparing lectures, many biologists spend quite a bit of time writing grant proposals to fund their research and evaluating proposals submitted by colleagues. A research proposal is unquestionably an argument; success depends on our ability to convince a panel of other biologists that what we wish to do is worth doing, that we are capable of doing it, that we can interpret the results correctly, that the work cannot be done without the funds requested, and that the amount of funding requested is appropriate for the research planned. Research money is not plentiful. Even well-written proposals have a difficult time; poorly written proposals generally don't stand a chance.

When we are not writing grant proposals or lectures, we are often preparing the results of our research for publication or for presentation at meetings. Research articles are really just laboratory reports based on data collected over a period much longer than the typical laboratory session. In research articles, as in laboratory reports, the goal is to make a strong case for doing the research, to present data clearly, and to interpret those data thoroughly and convincingly in the context of previous work and basic biological principles. Preparing research reports typically involves the following steps:

1. Organizing and analyzing the data
2. Preparing a first draft of the article (following the procedures outlined in Chapter 9 of this book)
3. Revising and reprinting the paper
4. Asking one or more colleagues to read the paper critically
5. Revising the paper in accordance with the comments and suggestions of the readers

6. Reprinting and proofreading the paper
7. Sending the paper to the editor of the journal in which we would most like to see our work published

The editor then sends the manuscript out to be reviewed by 2 or 3 other biologists. Their comments, along with those of the editor, are then sent to the author, who must again rewrite the paper, often extensively. The editor may then accept or reject the revised manuscript, or the editor may request that it be rewritten again before publication.

Oral presentations involve similar preparation. The data are organized and examined, a draft of the talk is prepared, feedback on the talk is solicited from colleagues, and the presentation is revised.

College and university biologists also write about you. Letters of recommendation are especially troublesome for us, because they are so important to you. Like a good laboratory report, literature review, essay, or term paper, a letter of recommendation must be written clearly, developed logically, and proofread carefully. It must also support all statements of opinion with facts or examples if it is to argue convincingly on your behalf and help get you where you want to go.

And then, there are progress reports to write, and committee reports, and internal memoranda. All this writing involves thinking, organizing, nailing down convincing arguments on paper, revising, retyping, and proofreading.

Clearly, being able to write effectively will help advance your career. Clear, concise, logical writing is an important tool of the biologist's trade: Learning how to write well is at least as important as learning how to use a balance, extract DNA, use a taxonomic key, measure a nerve impulse, run an electrophoretic gel, or clone a gene. And unlike these rather specialized laboratory techniques, **mastering the art of effective writing will reward you regardless of the field in which you eventually find yourself**.

In preparing the cover letter that accompanies a job application, for example, you are again building an argument: You are trying to convince someone that you understand the position you are applying for, that you have the skills to do the job well, and even that you want to do the job well. Similarly, in constructing a business plan, you must write clearly, concisely, and convincingly if you are to get your project funded. The fact that you may not become a biologist is no reason to cheat yourself out of the opportunity to become an effective writer. Remember this: While you're in college, you have a captive audience; **once you graduate, though, nobody *has* to read anything you write ever again**.

THE KEYS TO SUCCESS

It's always easier to learn something than to use what you've learned.
CHAIM POTOK, *THE PROMISE*

There is no easy way to learn to write well in biology or in any other field. It helps to read a lot of good writing, and not just in biology. Whenever you read anything that seems especially clear or easy to follow, examine that writing carefully to see what made it work so well for you. Reading well-written sentences aloud can also help plant good patterns in your brain. But mostly you just have to work hard at writing—and keep working hard at it, draft after draft, assignment after assignment. That will be much easier to do if you have something in mind that you actually want to say. Much of this book is about how to get to that point.

All good writing involves 2 struggles: the struggle for understanding, and the struggle to communicate that understanding to readers. Like the making of omelettes or crepes, the skill improves with practice. There are no shortcuts, and there is no simple formula that can be learned and then applied mindlessly to all future assignments. Every new piece of writing has to be thought about anew. Being aware of certain key principles, however, will ease the way considerably. Each of the following rules is discussed more fully in later chapters (note the relevant page numbers). This listing is worth reading at the start of each semester, or whenever you begin a new assignment.

Eleven Major Rules for Preparing a First Draft

1. **Work to understand your sources (pp. 34–44).** The only things we ever really learn are things we teach ourselves. When writing laboratory reports, spend time wrestling with your data until you are convinced you see the significance of what you have done. When taking notes from books or research articles, reread sentences you don't understand, and look up any words that puzzle you. Take notes in your own words; extensive copying or paraphrasing usually means that you do not yet understand the material well enough to be writing about it. Too few students take this struggle for understanding seriously enough, but all good scientific writing begins here. You can excel—in college and in life after college—by being one of the few who meet this challenge head on. Do not be embarrassed to admit—to yourself or to others—that you do not understand something after working at it for a while. Talk about the material with other students or with your instructor. If you don't commit yourself to winning the struggle for understanding,

you will either end up with nothing to say or what you do say will be wrong. In both cases, you will produce nothing worth reading.

2. **Don't quote from your sources.** Direct quotations rarely appear in the formal biological literature. Describe what others have done and what they have found, but do so in your own words. Consider this sentence:

    Shell adequacy was measured by the "shell adequacy index," defined by Vance (1972) as "the ratio of the weight of the hermit crab for which the shell was of preferred size to the actual weight of the hermit crab examined."

    When I see writing like this, I suspect that the writer did not understand the material being quoted; when you understand something thoroughly, you should be able to explain it in your own words. Perhaps I'm being unfair: Maybe this student just couldn't think of how to explain this better than the author already did. But if the student were explaining the shell adequacy index to a fellow student, would he or she have used that wording? I don't think so. Always think of yourself as explaining things to others, and do so using your own words. With practice and conscientious effort, you will find yourself capable of presenting facts and ideas in perfectly fine prose of your own devising.

3. **Don't plagiarize. See my detailed discussion later in this chapter (pp. 13–15).** We all build on the work and ideas of others. Whenever you restate another writer's ideas or interpretations, you must do so in your own words and credit your source explicitly. You don't lose face by crediting your sources. To the contrary, you demonstrate to readers your growing mastery of the literature. Note, too, that simply changing a few words or changing the order of a few words in a sentence or a paragraph is still plagiarism. Plagiarism is one of the most serious crimes in academia: It can get you expelled from college or cost you a career later.

4. **Think about where you are going before you begin to write (pp. 133–135, 223–228).** Much of the real work of writing is in the thinking that must precede each draft. Effective writing is like effective sailing; you must take the time to plot your course before getting too far from port. Your ideas about where you are going and how best to get there may very well change as you continue to work with and revise your paper, because the act of writing invariably clarifies your thinking and often brings entirely new ideas into focus. Nevertheless, you must have some plan in mind even when you begin to write your first draft. This plan

evolves from thoughtful consideration of your notes. Think first, then write; thoughtful revision follows.

Some people find it helpful to think at a keyboard or with a pen in hand, letting their thoughts tumble onto the paper. Others prefer to think "inside," writing only after their thoughts have come together into a coherent pattern. Either way, the hard work of thinking must not be avoided. If you still don't know where you are heading when you sit down to write that last draft of your paper, you certainly won't get to your destination smoothly, and you may well not get there at all. Almost certainly, your readers will not get there.

5. **Practice summarizing information (p. 42).** The longer I work with writing issues, the more I realize the central importance of being able to summarize information effectively. If you can't summarize the results of one research paper in your own words, you can't possibly see the relationship between 3 or 4 such papers; summary is an essential prelude to synthesis. The more practice you get summarizing information in your own words, the better. After you hear a lecture, take 10 minutes to summarize the major points in your notebook. After you see a movie or read a book, a short story, or even a newspaper article, try writing a one-paragraph summary every now and then. From time to time, after reading even a single paragraph of something, try writing a one-sentence summary of that paragraph (see p. xx for an example). The ability to summarize is an underappreciated, largely neglected, but essential skill for professional life.
6. **Write to illuminate, not to impress (p. 94, 206).** Use the simplest words and the simplest phrasing consistent with that goal. Avoid acronyms, and define all specialized terminology. In general, if a term was recently new to you, it should be defined in your writing. And if you can talk about "zones of polarizing activity" instead of "ZPAs," please do so. Your goal should be to communicate; why deliberately exclude potentially interested readers by trying to sound "scientific"? Don't try to impress readers with big words and a technical vocabulary; focus instead on getting your point across.
7. **Write for your classmates and for your future self (pp. 88, 206).** It is difficult to write effectively unless you have a suitable audience in mind. It helps to write papers that you can imagine being interesting to and understood by your fellow students. You should also prepare your assignments so that they will remain meaningful to *you* should you read them far in the future, long after you have forgotten the details of coursework completed or experiments performed. Addressing these 2 audiences—your fellow students and your future self—should help you write both clearly and convincingly.

8. **Support all statements of fact and opinion with evidence (pp. 32, 64–66, 69–73, 138–139, 188–191).** Remember, you are making arguments. In any argument, a statement of fact or opinion becomes convincing to the critical reader only when that statement is supported by evidence or explanation; provide it. You might, for instance, write the following:

   > Among the vertebrates, the development of sperm is triggered by the release of the hormone testosterone (Gilbert, 2003).

   In this case, a statement of fact is supported by **reference to a book** written by Scott Gilbert in 2003. Note that an author's first name is never included in the citation. In the following example, a statement is backed up by **reference to the writer's own data**:

   > Some wavelengths of light were more effective than others in promoting photosynthesis. For example, plants produced oxygen nearly 4 times faster when exposed to light of 650 nm* than when exposed to a wavelength of 550 nm (Figure 2).

   Statements can also be supported by **reference to the results of statistical analyses**, as in the following example:

   > The blue mussels produced significantly thicker shells in the presence of crustacean predators ($t = 4.65$; *d.f.* = 23; $P < 0.01$).

   Here, the statement is supported by the results of a *t*-test. The meaning of the items in parentheses is further explained in Chapter 4.
9. **Always distinguish fact from possibility.** In the course of examining your data or reading your notes, you may form an opinion. This is splendid. But you must be careful not to state your opinion as though it were fact. "The members of species *X* lack the ability to respond to sucrose" is a statement of fact and must be supported with a reference. "Our data suggest that adults of species *X* lack the ability to respond to sucrose" or "Adults of species *X* seem unable to respond to sucrose" expresses your opinion and should be supported by drawing the reader's attention to key elements of your data set. Similarly, consider the following statement:

   > The data suggest that "vestigial wings" is an autosomal recessive trait whereas "carnation eyes" is a sex-linked recessive trait.

---

*nm = nanometers; that is, 10^{-9} meters

Adding that one phrase, "the data suggest," makes all the difference. Basing an opinion solely on his or her own data, the writer would be on far shakier ground beginning the sentence with "'Vestigial wings' is an autosomal recessive trait."

10. **Allow time for revision (pp. 80–82).** Accurate, concise, successfully persuasive communication is not easily achieved, and few of us come close in a first or even a second draft. Although the act of writing can itself help clarify your thinking, it is important to step away from the work and reread it with a fresh eye before making revisions; a "revision" is, after all, a re-vision: another look at what you have written. This second (or third, or fourth) look allows you more easily to see if you *have* said what you had hoped to say, and whether you have guided the reader from point to point as masterfully as you had intended. Remember, you are constructing an argument. It takes thoughtful revision to make any argument fully convincing. Start writing assignments as soon as possible after receiving them, and always allow at least a few days between the penultimate and final drafts. If you follow this advice, the quality of what you submit will improve dramatically, as will the quality of what you learn from the assignment.
11. **Back up your drafts every few minutes** on your hard drive. At the end of each session, make another backup copy on a flash drive or send a copy of the file to yourself as an email attachment.

Six Major Rules for Developing Your Final Draft

Once you have nailed together the basic framework of your presentation or argument, it is time to tighten the construction.

12. **Stick to the point (p. 136, 207, 234).** Delete any irrelevant information, no matter how interesting it is to you. Snip it out and put it away in a safe place for later use if you wish, but don't let asides interrupt the flow of your writing.
13. **Say exactly what you mean (pp. 85–93).** Words are tricky; if they don't end up in the right places, they can add considerable ambiguity to your sentences. For example, "I saw 3 squid SCUBA diving last Thursday" conjures up a very interesting image. Don't make readers guess what you're trying to say; they may guess incorrectly. Good scientific writing is precise. Sloppy writing often implies sloppy thinking. Figure out exactly

what you mean to say, and be sure you then say what you mean. It often helps to read aloud what you have written and to listen carefully to what you say as you read.

14. **Never make the reader back up (pp. 102–105).** You should try to take readers by the nose in your first paragraph and lead them through to the end, line by line, paragraph by paragraph. Link your sentences carefully, using transitional words, such as *therefore* or *in contrast*, or by repeating key words so that a clear argument is developed logically. Remind the reader of what has come before, as in the following example:

    In saturated air (100% relative humidity), the worms lost about 20% of their initial body weight during the first 20 hours but were then able to prevent further dehydration. In contrast, worms maintained in air of 70–80% relative humidity dehydrated far more rapidly, losing 63% of their total body water content in 24 hours. As a consequence of this rapid dehydration, most worms died within the 24-hour period.

    Note that the second and third sentences in this example begin with transitions ("In contrast," "As a consequence of"), thus continuing and developing the thought initiated in the preceding sentences. A far less satisfactory last sentence might read "Most of these animals died within the 24-hour period."

    Link your paragraphs in the same way, using transitions to continue the progression of a thought, reminding the readers periodically of what they have already read.

    Avoid casual use of the words *it*, *they*, and *their*. For example, the sentence "It can be altered by several environmental factors" forces the reader to go back to the preceding sentence, or perhaps even to the previous paragraph, to find out what *it* is. Changing the sentence to "The rate of population growth can be altered by several environmental factors" solves the problem. Here is another example:

    Our results were based upon observations of short-term changes in behavior. They showed that feeding rates did not vary with the size of the caterpillar.

    In this example, the word *they* could refer to "results," "observations," or "changes in behavior." Granted, the reader can back up and figure out what "they" are, but you should work to avoid the "You know what I mean" syndrome. Changing "they" in the second sentence to "These results" avoids the ambiguity and keeps the reader moving effortlessly in the right direction.

Do not be afraid to repeat a word or phrase used in a preceding sentence; if it is the right word and avoids ambiguity, use it. Repetition can be an effective way to keep readers moving forward.

15. **Don't make readers work harder than they have to (pp. 92–96, 188–190).** If there is interpreting to be done, you must be the one to do it. For example, never write something like:

    The difference in absorption rates is quite clearly shown in Table 1.

    Such a statement puts the burden of effort on the reader. Instead, write something like:

    Clearly, alcohol was more readily absorbed into the bloodstream from distilled beverages than from brewed beverages (Table 1).

    Readers now know exactly what you have in mind and can examine Table 1 to see if they agree with you.

16. **Be concise (pp. 96–102).** Give all the necessary information, but avoid using more words than you need for the job at hand. By being concise, your writing will gain in clarity. Why say:

    Our results were based upon observations of short-term changes in behavior. These results showed that feeding rates did not vary with the size of the caterpillar.

    when you can say:

    Our observations of short-term changes in behavior indicate that feeding rates did not vary with the size of the caterpillar.

    In fact, you might be even better off with the following sentence:

    Feeding rates did not vary with caterpillar size.

    With this modified sentence, nearly 70% of the words in the first effort have been eliminated without any loss of content. Cutting out extra words means you will have less to type; you'll have your paper finished that much sooner. Finally, your readers can digest the paper more easily, reading it with pleasure rather than with impatience.

17. **Don't be teleological (p. 106).** That is, don't attribute a sense of purpose to other living things, especially when discussing

evolution. Giraffes did not evolve long necks "in order to reach the leaves of tall trees." Birds did not evolve nest-building behavior "in order to protect their young." Insects did not evolve wings "in order to fly." Plants did not evolve flowers "in order to attract bees for pollination." Natural selection operates through a process of differential survival and reproduction, not with intent. Long necks, complex behavior, and other such genetically determined characteristics may well have given some organisms an advantage in surviving and reproducing that was unavailable to individuals lacking those traits, but this does not mean that any of these characteristics were deliberately evolved in order to achieve something.

Organisms do not evolve structures, physiological adaptations, or behaviors out of desire. Appropriate genetic combinations must always arise by random genetic events—by chance—before selection can operate. Even then, selection is imposed on the individual by its surroundings and, in that sense, is a passive process; natural selection never involves conscious, deliberate choice. Don't write, "Insects may have evolved flight in order to escape predators." Instead, write, "Flight among insects may have been selected for by predation."

Nine Finer Points: The Easy Stuff

18. **Abbreviate units of measurement that are preceded by numbers.** Do not put periods after unit symbols, and always use the same symbol for all values regardless of quantity: 1 mm (millimeter), 50 mm; 1 hr (hour), 50 hr; 1 g (gram), 454 g.
19. **Always underline or italicize species names, as in *Homo sapiens*.** Note also that the generic name (*Homo*) is capitalized whereas the specific name (*sapiens*) is not. Once you have given the full name of the organism in your paper, the generic name can be abbreviated; *Homo sapiens*, for example, becomes *H. sapiens*. There is no other acceptable way to abbreviate species names. In particular, it is not permissible to refer to an animal using only the generic name, because most genera include many species. (Note that the plural of *genus* is *genera,* not *genuses.*)
20. **Don't use formal scientific names to refer to individuals of a species.** For example, instead of writing that "*Chromys ludovicianus* is often considered an ecosystem engineer because it modifies its surroundings so extensively through its feeding and

burrowing activities (Coppock et al., 1983)," write "Individuals of *Chromys ludovicianus* are often considered ecosystem engineers ...," or use the common name for the species like this: "Black-tailed prairie dogs (*Cynomys ludovicianus*) are often considered ecosystem engineers because they modify their surroundings so extensively ..."

21. **Do not capitalize common names.** Examples include monarch butterfly, lowland gorillas, pygmy octopus, and fruit fly.
22. **When listing references at the end of a sentence, put the period after the references.** For example, "Most of what we currently know about how animals orient to magnetic fields is based on studies of vertebrates (Able and Able, 1995; Phillips, 1996)."
23. **Capitalize the names of taxonomic groups (clades) above the level of genus, but not the names of the taxonomic categories themselves.** For example, insects belong to the phylum Arthropoda and the class Insecta. Do not capitalize informal names of animals: Insects are arthropods, members of the phylum Arthropoda.
24. **Remember that the word *data* is plural.** The singular is *datum*, a word rarely used in biological writing. "The data are lovely" (not "The data is lovely"). "These data show some surprising trends" (not "This data shows some surprising trends"). You would not say, "My feet is very large"; treat *data* with the same respect.
25. **Pay attention to form and format: Appearances can be deceiving.** Your papers and reports should give the impression that you took the assignment seriously, that you are proud of the result, and that you welcome constructive criticism of your work. Type or computer-print your papers whenever possible; use only one side of each page. Leave margins of about an inch and a half on the left and right sides of the page and about an inch at the top and bottom of each page. Double-space your typing so that your instructor can easily make comments on your paper. Make corrections neatly. Never underestimate the subjective element in grading.
26. **Put your name and the date at the top of each assignment, and number all pages.** Pages should be numbered so that readers can tell immediately if a page is missing or out of order and can easily point out problems on particular pages ("In the middle of page 7, you imply that ..."). Remarkably, most word processing programs do not automatically number pages for you; you must tell the program to insert the page numbers.

The Annoying but Essential Last Pass

27. **Proofread.** None of us likes to proofread, even though it is a crucial part of the writing process. By the time we have arrived at this point in the project, we have put in a considerable amount of work and are certain we have done the job correctly. Who wants to read the paper yet another time? Moreover, finding an error means having to make a correction. But put yourself in the position of your instructor, who must read perhaps 100 or more papers and reports each term. He or she starts off on your side, wanting to see you earn a good grade. Similarly, a reviewer or editor of scientific research manuscripts starts off wanting to see the paper under consideration get published. A sloppy paper—for example, one with many typographical errors—can lose you a considerable amount of goodwill as a student and later as a practicing scientist. For one thing, sloppy writing may suggest to the reader that you are equally sloppy in your work and in your thinking, or that you take little pride in your own efforts. Furthermore, failure to proofread your paper and to make the required corrections implies that you don't value the reader's time. That is not a flattering message to send, nor is it a particularly wise one. Never forget: There is often a subjective element to grading and to decisions about the fate of manuscripts and grant proposals. For all these reasons, shoddily prepared material can easily lower a grade, damage a writer's credibility, reduce the likelihood that a manuscript will be accepted for publication or that a grant proposal will be funded, or cost an applicant a job or admission to professional or graduate school. Why put yourself in such jeopardy in order to save a mere half-hour? **Turn in a piece of work that you are proud to have produced**.

AVOIDING PLAGIARISM

The paper or report you submit for evaluation must be original: It must be *your* work. **Submitting anyone else's work under your own name is plagiarism**, even if you alter some words or reorder some sentences. **Presenting someone else's thoughts or ideas as your own is also plagiarism**. Consider the following 2 paragraphs:

Smith (1991) suggests that this discrepancy in feeding rates may reflect differences in light intensities used in the two experiments. Jones (1994), however, found that light intensity did not influence the feeding rates of these animals and suggested that

> the rate differences instead reflect differences in the density at which the animals were held during the two experiments.

> This discrepancy in feeding rates might reflect differences in light intensities. Jones (1994), however, found that light level did not influence feeding rates. Perhaps the difference in rates reflects differences in the density at which the animals were held during the two experiments.

The first example is fine: Every idea is clearly associated with its source. In the second example, however, the writer takes credit for the ideas of Smith and Jones; the writer has plagiarized.

As another example, consider this student's summary sentence based on the Rachel Carson passage presented on p. 42:

> The world that supports the animal life on planet earth is made up of water, soil, and a green mantle of plants.

That's plagiarism in action. This is not:

> Rachel Carson (1962) stresses that animal life on our planet could not exist without plants.

The second student has *processed* information; the first student has not.

Here is another example, one that might surprise you. Suppose you hear a talk by an outside speaker on campus, hosted by the Biology Department. You go because the speaker's research on reproductive isolation in the European corn borer relates to the topic of a paper or laboratory report that you are writing. Time well spent: You leave the talk with several pages of notes, and you even ask the speaker a few questions after the talk is over.

In writing your paper or report, you borrow a number of the speaker's ideas but don't attribute them to the speaker. Instead, you present the ideas as your own. You have plagiarized.

What you have done is not only immoral; it's also self-defeating. If you had credited the speaker with those ideas you would have impressed your instructor fabulously—an undergraduate actually attends a departmental seminar without being required to do so and pays close enough attention to incorporate some of the speaker's information in the paper. You might be able to write something like this: "As Professor Dopman noted in his talk (October 12, 2011), the release of pheromones by females does not always result in courtship by males," and expand that thought in the next sentence or two. That would be very impressive. Crediting ideas and working effectively with other people's ideas show your growing command of the field. That's a GOOD thing.

But here's some bad news: **Plagiarism is theft**. It is one of the most serious offenses that can be committed in academia, where original thought is the major product of one's work—often months, sometimes years of physical and mental work. At the very least, an act of plagiarism will result in an F on the assignment or for the entire course. Repeated plagiarism (but sometimes even a single offense) can get you expelled from college.

And now for the really bad news: **Plagiarism is getting easier to detect**. Computer programs designed to detect plagiarism are now being used at more and more colleges and universities (and by more scientific journal editors). Such programs, for example, Turnitin, search enormous databases that include millions of pages from books and journals, millions of Web sites, and tens of millions of papers previously submitted by college students across the country.

Doing a Google search on the phrase "avoiding plagiarism" brought back more than 313,000 entries, so plagiarism is clearly a problem of great concern to many people. Intentional plagiarism (e.g., copying text directly from a Web site or putting your own name on a paper written by someone else) is easy to avoid: Just don't do it! But some plagiarism is unintentional; how do you avoid that? One approach is to read widely. If all you know about a particular topic is what you have read in your laboratory manual or textbook, your options for original thinking will be limited. **The more you read, the more you will have to draw on in expressing your own thoughts.** You also need to give yourself time to digest what you've read before you can begin putting the material together in an original way, in your own voice and with a clear sense of direction. If you wait until the last weekend to start reading, or you read without thinking, it's hard to avoid plagiarism in writing your paper.

Plagiarism can also occur unintentionally through bad note-taking practices. **Take notes in ways that minimize the likelihood of plagiarism**, as discussed in Chapter 3 (pp. 42–44). An added benefit of developing good note-taking techniques is that you will come away with a much greater understanding of what you read, and you will have much more substantive things to talk about and write about.

Another way to help avoid unintentional plagiarism is to write your first draft without looking at your notes. If you can't do this, you're probably not ready to write anyway.

ON USING COMPUTERS IN WRITING

With computers, perfection is within your immediate grasp. Instructors therefore find it increasingly annoying when students turn in computer-printed reports that are carelessly written and not proofread. Word processing has become a two-edged sword.

Let me also warn you about what you cannot expect a computer to do for you. Computers can do little to help you in that all-important first struggle—the struggle for understanding. Neither (unfortunately) can they think, organize, or revise for you. Computerized spelling checkers can catch some typographical and spelling errors, but you cannot expect them to catch all of your mistakes. Biology is a field with much specialized terminology, much of which is of no use to nonbiologists; these terms therefore do not find their way into the dictionaries that accompany computerized spelling programs. Although you can easily add words to the computer's dictionary, the terminology in your papers will be changing with every new assignment; many of the words you add for today's assignment will probably not be used in next week's assignment. Moreover, a spelling-checker program will not distinguish between *to* and *too*, *there* and *their*, or *it's* and *its*, and the program will miss typographical errors that are real words. Suppose, for example, that you typed *an* when you intended to type *and*, or you typed *or* when you should have typed *of*, or you typed *rat* instead of *rate*. Grammar-checker programs also will not catch every error, and even when they recognize mistakes they do not always suggest the proper correction. By all means, use spelling- and grammar-checker programs for a first pass, but then use your own sharp eyes and keen intellect—moving word by word and sentence by sentence—to complete the necessary process of proofreading your work.

TECHNOLOGY TIP 1

Using Shortcuts and Autocorrect

The field of biology contains much specialized terminology; a number of abbreviations that require superscripts or subscripts; and many long, tongue-twisting species names (which always need to be italicized—Rule 19). Keyboard shortcuts and the AutoCorrect feature found in most word processing programs can be very helpful in dealing with these issues. For example, here are things you can do in Microsoft Word:

1. **For italicizing**, instead of highlighting and then clicking on the *I* symbol in the bottom line of the Word menu system (to the right of the **B** symbol, for boldfacing), press the Ctrl key and type an "i" before and after the word(s) that you want italicized; leave no space between the "i" and the letters of the word. You can **boldface** terms in a similar way, by typing Ctrl-b before and after the word.

2. **For superscripts**, hit Ctrl Shift = and then type the text that you would like superscripted. Repeat the process to exit superscript mode.
3. **For subscripts**, hit Ctrl = and then type the text you would like subscripted. Repeat the process to exit subscript mode.
4. **For long words**, such as complicated chemical names (e.g., fructose-1,6-diphosphate) that you need to type repeatedly, use the **AutoCorrect** feature of Word as follows:
   - Go to Tools in the tool bar menus and select AutoCorrect Options
   - Decide on an abbreviation for the word you will need to type repeatedly. For "fructose-1,6-diphosphate," for example, you could use the abbreviation "frc." Type the abbreviation into the Replace space. Be sure that your abbreviation is not a combination of letters that appears frequently in other words; never use "the," for example.
   - Type the full version (e.g., "fructose-1,6-diphosphate" [without the quotation marks]) into the With space. **Double-check your spelling before proceeding!**
   - Click Add, and then OK. Now, every time that you type the letters "frc" and hit the space bar, Word will automatically "correct" this to the name of the sugar that you programmed in.
5. For **italicized words that you will use repeatedly, such as the formal name of the sea urchin genus** *Strongylocentrotus*, do the following:
   - Type the complete word once in your text, italicize it, and highlight just that word.
   - Click on Tools and then AutoCorrect. Your highlighted word will appear in the With space.
   - Enter a simple abbreviation for that word (e.g., Sts) in the Replace space.
   - Be sure that the "Formatted text" circle is selected. Click Add and then OK.
   - From now on, type Sts and you will instantly see *Strongylocentrotus* on your screen, spelled correctly and beautifully italicized, as soon you hit the space bar.

(*Continued*)

6. For automatic formatting of **subscripts and superscripts**, as in the expression K_m, associated with enzyme kinetics, do the following:

   - Type the letters you want (in this case, Km) into your text.
   - Turn the "m" in our example into a subscript using the Format and then Font tools in the Tool Bar (check the subscript box, in this case) or by using one of the shortcuts mentioned above.
   - Click Tools and then AutoCorrect. Your highlighted letters will appear in the With space.
   - Type Km into the Replace space.
   - Click Add, and then click OK. From now on, whenever you type Km and hit the space bar, you will automatically see K_m on your screen.
   - You can play the same trick with **other specialized symbols**, such as μm (micrometer), °C (degrees Centigrade), and ‰ (parts per thousand, referring to salinity). AutoCorrect can be a writer's best friend!
   - Caution: Note that if you program "um" as a default abbreviation for μm, you may get "μmbrella" whenever you type "umbrella." The letters "umc" are a much safer choice. Choose your codes carefully, and keep a list of the codes that you have programmed.

ON USING COMPUTERS FOR DATA STORAGE, ANALYSIS, AND PRESENTATION

In addition to their use as word processors, computers are also used by many biologists for storing and retrieving literature references and for storing and analyzing data. Some data sets, particularly in ecology, are often too complex to analyze any other way. But undergraduate biology majors will probably find that a set of note cards and a $30 scientific calculator will be perfectly adequate for what they will be asked to do in most courses.

On the other hand, computers can be a real help in preparing your graphs and tables, as discussed in Chapter 9. Whether prepared on computer or by hand, all graphs must be carefully planned, sensible, and neatly executed.

Computers are now routinely used for formal presentations in classes and at conferences. With appropriate software, all of your visual displays can now be created on and displayed from a laptop computer, in brilliant color, with palm tree backgrounds, and even with titles and pointers sliding in from the sidelines at the push of a button as you talk. But **the substance of what you say and the extent to which you communicate that substance to your audience are what really matter**. With good planning, you can give a perfectly wonderful talk using the chalkboard or simple overheads in black and white. Effective communication should always be your primary goal; don't try to hide behind the technology. If you aren't careful, a high-tech presentation can be a barrier to effective communication, as discussed in Chapter 11.

SUMMARY

1. Acknowledge the struggle for understanding, and work to emerge victorious. Read with a critical, questioning eye (p. 4, 34–43, 126–127, 133–135, 223–224).
2. Think about where you are going before you begin to write, while you write, and while you revise (p. 5–6, 82–86, 102–104, 133–134, 223–224).
3. Never miss an opportunity to practice summarizing information in your own words (p. 6, 39–43).
4. Write to illuminate, not to impress (p. 6, 81, 88– 89, 206, 235, 240).
5. Write for an appropriate audience—for example, your classmates and your future self (p. 6, 88–89).
6. Back up all statements of fact or opinion (p. 7, 69–73, 188–190, 206).
7. Always distinguish fact from possibility (p. 9–10, 206).
8. Don't quote, and don't plagiarize (p. 5, 13–15, 43). Be careful not to take credit for the work or ideas of others.
9. Allow adequate time for revision (p. 8, 80–81).
10. Stick to the point (p. 8, 136, 207, 234).
11. Say exactly what you mean (p. 8–9, 85–93).
12. Never make the reader back up (p. 9–10, 102–105).

13. Be concise: Avoid unnecessary words, unnecessary jargon, weak verbs, and unnecessary prepositions (p. 16, 92–94, 96–102, 206).
14. Avoid teleology (p. 10–11, 106).
15. Save your computer work frequently—at least every few paragraphs—and always make a backup copy before ending a session (p. 8).
16. Proofread all work before turning it in, and keep a copy for yourself.
17. Underline or italicize the scientific names of species.
18. Remember the word *data* is plural, not singular.
19. Make your papers neat in appearance. Double-space all work, and leave margins for the instructor's comments and suggestions.
20. Put your name and the date at the top of each assignment, and number all pages.
21. When giving talks, never let style and technology become more important than the substance of what you are presenting (p. 19, 241).

2

LOCATING USEFUL SOURCES

You may think of [investigation] as a dull, plodding effort involving very little thinking of any kind, let alone creative thinking. In part, that's right. The way many people actually carry out their investigation involves little or no thinking—which is why their investigation is so often unproductive.

VINCENT RUGGIERO

What Lies Ahead? In This Chapter, You Will Learn

- The difference between the primary literature and the secondary literature
- That references found in refereed research articles can be your best source of further information, once you've located a good key article from the primary literature
- How to use ISI *Science Citation Index*, *Medline*, and other databases effectively
- How to conduct efficient and effective searches on the Internet and subscription databases
- That finding and reading carefully a small number of especially relevant sources is usually far more productive than collecting hundreds of random references that you don't have time to read

For most assignments in upper-level biology courses, you will be asked to go beyond the factual foundation of a field and read the **primary literature**, which presents the results of original studies and includes detailed information about how those studies were conducted. Articles encompassing this primary literature are written by the people who did the research, and are published in formal research journals such as *Science*, *Ecology*, *Genome Biology*, *Cell*, *Biological Bulletin*, *BMC Neuroscience, and PLoS Biology.* The articles are published only after being reviewed ("refereed") and evaluated by other scientists in a process called "peer review."

But you can't search effectively for useful primary sources until you have a pretty good idea of what you're looking for. Begin by reading the relevant portions of your class notes and relevant documents in the **secondary literature**, which gives someone else's summaries, interpretations, and evaluations of the primary literature. Textbooks, symposium volumes, review papers (as in *Scientific American* or *Quarterly Review of Biology*), newspaper and magazine articles, encyclopedia entries, and Wikipedia articles (see caution about Wikipedia later in this chapter, p. 27!) are all examples of this secondary literature. Armed with sufficient background, you will now be ready to locate—and delve into—the more specialized material found in the primary literature, the original research papers that are published in peer-reviewed journals.

When seeking relevant books for background, you may have to become a little devious before you can convince your library's online system to satisfy your request for information. Suppose, for example, that you wish to find material on reptilian respiratory mechanisms. You might try, to no avail, looking under Respiration or Reptiles, but looking under Physiology or Comparative Physiology will probably pay off. Similarly, in researching the topic of annelid locomotion, you might try, unprofitably, searching under Annelids, Locomotion, or Worms. Looking under Invertebrate Zoology, however, will probably turn up something useful. If at first you don't succeed ... ask a reference librarian.

EASY WAYS TO ACCESS THE PRIMARY LITERATURE

The references given in textbooks and review articles often provide good access to the primary literature. Especially good sources of reviews include *Integrative and Comparative Biology, Biological Reviews, BioScience, Scientific American, Quarterly Review of Biology*, and the *Annual Review* series (e.g., *Annual Review of Ecology and Systematics, Annual Review of Genetics*). Once you locate a review on the topic you wish to explore, look carefully at the citations used to support statements of particular interest. For example, consider this brief excerpt from a paper published in *Biological Reviews**:

> Chimpanzees perhaps make the most frequent and diverse use of tools in the wild (Goodall, 1973; Tomasello, 1990). Chimpanzees living in one region of West Africa use a pair of stones (a hammer and an

*Miklósi, A. 1999. The ethnological analysis of imitation. *Biol. Rev*. 74: 347–374.

> anvil) to open oil-palm nuts (e.g., Sugiyama and Koman, 1979; Boesch and Boesch, 1983). The nut-cracking process is thought to be one of the most difficult learned tasks performed by any animal in the wild. Chimpanzees place a nut on a suitable stone or sometimes a root (the 'anvil') and hit it with a carefully chosen stone or piece of wood ('the hammer'). Chimpanzees not only transport hammers; some wooden hammers are made by the chimpanzees (Boesch and Boesch, 1990). One possibility is that the chimpanzees understand the logical structure of the task (Byrne, 1994) and the behaviour sequence (taking the stone into the hand, putting the nut on the …).

A student interested in the manufacture of tools by animals might add Boesch and Boesch (1990) to his or her list of papers to read. The title and complete reference for that paper will be found at the end of the article, in the Literature Cited section.

It is also profitable to browse through recent issues of specialized journals in the primary literature relevant to your topic. Ask your instructor to name a few journals worth looking at, or ask your reference librarian for help in finding core journals in your research area. **If you find an appropriate article in the recent literature, examine the literature citations at the end of that article for additional references**, particularly those used to support statements of special interest made in the article's text. This is an especially easy and efficient way to accumulate research articles; the yield of good references is usually high for the amount of time invested.

Note that the distinction between primary and secondary literature can be blurred at times, depending on how you are using the primary literature that you read. If you are using just the Introduction section of a research paper for background or to locate additional research articles on a topic, for example, then that research article is serving as the secondary literature for you: You are using it as a review of previous research. So the secondary literature is always just that, but the research literature itself can be both primary and secondary, depending on how you use it.

USING INDEXES

Hundreds of thousands of research papers are published in biology every year, in many thousands of journals and conference proceedings. Thumbing through journals at random is not an efficient—or even feasible—way to conduct a thorough search of the literature. **For thorough and efficient searches on particular topics, use one or more indexes.** Using electronic indexes, computers can now examine about 100 years of source material and then give you a list of all references relevant to the information you provided, in a matter of seconds. Knowing how to input your search terms in a way the computer "understands" is the key to conducting an effective search.

The most widely used indexes are ISI *Science Citation Index, Biological Abstracts, BIOSIS Previews*, and *Basic BIOSIS* (a subset of *Biological Abstracts*), all available through Thompson Reuter's Web of Knowledge and some other vendors. Another widely used index is *Medline,* produced by the U.S. National Library of Medicine (and freely available to all on PubMed). Most of these services provide abstracts of articles for free and free access to the full-text documents if your library has a subscription. Another long-running indexer is *Zoological Record*, published jointly by BIOSIS and the Zoological Society of London. The main distinguishing feature of *Zoological Record* is that it provides a means of looking up articles by taxonomic group, geographical location, or geological time period.

USING *SCIENCE CITATION INDEX*

Science Citation Index (one of the products available within the Thompson Reuters ISI Web of Knowledge) is a unique tool for locating recent references on a specific topic. To use this tool, you must first have located a key paper on the topic that was published at least 2 to 3 years ago. It might be a major review paper, for example, or a particularly interesting article in a major research journal. *Science Citation Index* allows you to find out what has happened subsequently, in that exact area, by listing more recent publications, with full citations, that include your key paper in their Literature Cited sections. A paper that cites your key paper is likely to have built on your topic of special interest. So, through this means, **you can quickly track a particular topic forward in time, even from a paper published many years ago**, to related papers published within the past year. Coverage extends back to 1899 (although many libraries do not have all years available), and about 18,000 new entries are added to the database each week, covering nearly 8,000 journals and 110,000 conference proceedings. *Science Citation Index* can also help you follow a topic even further back in time, because it will also list all the papers that are cited in each of the references in its database.

Science Citation Index also allows you to search the database by subject (e.g., parasite behavior), author name, or journal title. But the ability to track particular topics forward (and backward) in time from a key reference is what sets this service apart from others.

USING *CURRENT CONTENTS CONNECT*

Current Contents Connect essentially puts the best library in the world at your fingertips. Each weekly issue includes the complete table of contents for scientific journals published a few weeks earlier and even provides some information for certain articles about to be published. Nearly 8,500 journals are covered by the publication, along with more than 9,000 evaluated Web sites, so you are not likely to miss much of the relevant literature, no matter how meager the holdings in your institution's library. If you encounter a paper of particular interest while browsing the latest issue of *Current Contents*, you can try to find that article through your library's Web site or request a copy from your library's Document Delivery Service, if available. Failing that, **you can request a copy of the paper from its author**, because *Current Contents* lists the complete mailing addresses for the authors of each article. You can search the database by keyword, title, author, or journal name. The online database currently extends back to 1994.

USING *MEDLINE* AND OTHER DATABASES

Medline offers complete coverage of the biomedical literature back to 1946 and includes more than 11 million records. *Biological Abstracts* offers broader coverage of biological topics. *Basic BIOSIS* covers the primary research literature using a smaller subset of journals. *BioDigest* indexes popular articles in a variety of fields. Search in each of these indexes by entering a keyword (e.g., coughing), phrase (e.g., coughs induced by parasites), or Boolean search string (e.g., coughs AND parasite AND induced); the name of an author; or the title of a particular journal or article.

Other useful Web-based databases include the following:

- *Oceanic Abstracts*, which focuses on marine-related topics, including aquaculture, fisheries, and the effects of pollution
- *Agricola*, a database offered through the U.S. Department of Agriculture that includes more than 3 million records on agriculture and related topics from more 1,400 journals
- *Environment Complete* and *Environment Abstracts*, both of which provide multidisciplinary coverage of publications in the environmental sciences
- *Toxline*, which focuses on the toxicological literature
- *Pollution Abstracts*

- *Geobase*, which indexes the geological and ecological literature
- *Index to Scientific Reviews*, which indexes more than 30,000 newly published review articles each year
- *Annual Reviews*, which publishes a large variety of reviews each year in about 25 different areas of biology, including marine science, plant biology, ecology evolution and systematics, neuroscience, and genetics
- *CAB Abstracts*, which cover the fields of forestry, agriculture, animal health, human health and nutrition, and the conservation of natural resources
- *Scopus*, which covers more than 18,000 journals and 3 million conference papers in the health, agricultural, biological, and environmental sciences, including many published outside the United States (but which have abstracts in English). *Scopus* also includes 100% coverage of *Medline*. *Scirus* (p. 27) is included within *Scopus*. Like the ISI Web of Knowledge, you can access full-text articles through this site, for a fee.
- *JSTOR*, which archives complete runs of about 30 selected journals in the Ecology and Botany collection and another 16 journals in the Health and General Sciences collection
- *Science.gov*, which offers access to more than 200 million pages of official information from the Environmental Protection Agency, Department of Agriculture, and other U.S. government agencies

Please note that some databases now include references to papers published more than 100 years ago. In many areas of biology, **the older literature is a valuable and important resource** that should not be overlooked.

Your university librarian can tell you which of these services is available at your school, and how best to access each of them. You can access many of these databases even off campus by linking Google Scholar to your library's resources: Simply click the Scholar Preferences button and then enter the name of your college or university in the space labeled "Library Links."

PROWLING THE INTERNET

A phenomenal amount of information is out there in the ether, and the amount is growing at a truly astounding rate. Millions of Web sites and many billions of pages of information are now available on the Web, and thousands of new sites are added weekly.

This abundance of information creates both opportunities and problems. Prowling the Web can eat up your time the way a vacuum cleaner sucks up dirt. This is a rather good analogy, too, because much of what is on the Web is not worth reading. **Information presented in formal scientific journals and many books has gone through a rigorous peer-review process**; other scientists have evaluated and often shaped the information that is ultimately presented—and in some cases, even kept it from being published—but **most of the information available through the Internet has not been checked for accuracy**. The old adage "Don't believe everything you read" applies with a vengeance to most of what you will find on the Web. **Don't use any information presented on the Web that isn't offered by a recognized authority on the subject or that you can't verify using other sources.** The online encyclopedia Wikipedia can be an excellent source of background information and can often provide useful references for further consultation, but it cannot be cited as a valid, authoritative source because its content can be changed continuously and you do not know the competence, or the motives, of whoever is making those changes.

Choose your Web sites carefully. The most reliable information can be found through the Web sites of museums; recognized research organizations (e.g., the Marine Biological Laboratory in Woods Hole, Massachusetts); and major research universities, scientific societies, and government agencies (e.g., World Health Organization [WHO] and National Oceanic and Atmospheric Administration [NOAA]). Such sites will have .edu, .gov, or .org in their addresses; be careful with .org sites, however, as some will just be functioning as sales agents. (See Appendix C for additional information about evaluating Web sites.)

Many print journals offer full-text, online access through the Web. A listing of all journals currently offered online can be found at *SciCentral*. This site also provides access to numerous other biological databases, including organism-specific gene sequences and a range of standard protocols in molecular biology. In addition, the American Institute of Biological Sciences (AIBS) now provides Internet full-text access to more than 150 biological and environmental journals in a project called *BioOne*, first released in 2001. Every journal article in *BioOne* is linked to other relevant articles within the database, making it easy to assemble related references on particular topics.

Elsevier Science also launched an interesting new site in 2001 called *Scirus*, a search engine focused exclusively on science-related information. *Scirus* provides access not only to research article citations but also to the home pages of individual university and government researchers. Finally, Google offers an excellent search engine (*Google Scholar*) that includes scholarly writings in the sciences, along with relevant, selected references.

And now the catch: Indexes can lead you to useful references and, in many cases, will also provide for free the full abstract section of the published papers, but they may charge for access to the actual articles. You may be able to access the complete articles for free through your college library Web site, or a Google search might lead you to free access through the author's own Web site. Alternatively, the library might own a hard copy of the journals in which the articles were published. Some exclusively online journals also have appeared in recent years (e.g., *PLoS Biology, BMC Biology*) that provide free access to all of the articles that they publish. Such journals charge the authors a fee for publishing but, nevertheless, claim to maintain the same rigorous standards of peer review that characterize traditional "hard-copy" journals and so can be cited as you would any other authoritative source. BioMed Central publishes about 200 such online journals on an enormous range of topics, and the Directory of Open Access Journals (www.DOAJ.org) provides free, full-text access to carefully evaluated scientific and scholarly journals in a variety of subjects.

Check with your instructor or reference librarian to find out what is available and recommended at your school.

CONDUCTING WEB SEARCHES: DEVELOPING PRODUCTIVE SEARCH STRATEGIES

There are two other ways to search for information on the Web: directory searches and keyword searches. Some of the most useful search engines are listed in Table 1. To conduct a **directory search**, start with a main topic area (e.g., Biology), and then gradually narrow your search down step by step. From Biology, for example, you might click on the subtopics Neurobiology or BioDiversity, and then narrow the topic further within each of those subtopics. *Yahoo!* and the *Internet Public Library* are particularly good sites for conducting this type of directory search.

Alternatively, you might want to conduct a **keyword search**, in which you enter terms or phrases related to your subject of interest. The trick here is to limit your search so that you bring back a manageable number of references (see Technology Tip 2 on p. 30). Entering "Effects of parasites on salmon growth rates" for a search using *Scirus*, for example, brought up listings for almost 14,000 Web pages! That's ridiculous! If I were interested in finding out more about the effects of parasites on salmon growth rates, I might conduct an "advanced search" and ask the search engine to include just these terms: *parasites*, *salmon*, and *"growth rate."* (The quotation marks around *"growth rate"* tell the search engine

Table 1. Especially useful engines for searching the Web

Site	Address	Notes
Directory searches		
Yahoo!	http://www.yahoo.com	
Internet Public Library	http://www.ipl.org	See this site's Reference Center and Subject Collections (reviewed by librarians)
Keyword searches		
Google	http://www.google.com	
Google Scholar	http://scholar.google.com	Use "Advanced Search"
Scirus	http://www.scirus.com	Focuses exclusively on scientific material—use Advanced Search
Metasearch engine searches		
Surfwax	http://www.surfwax.com	Good for serious deep-digging
Ixquick	http://www.ixquick.com	

to look for that entire phrase, not just the isolated words *growth* and *rate*.) This search still brought in about 4,500 entries. Then, by excluding the terms *disease* and *nematode*, I was able to reduce the number of Web pages found to 938, certainly a more manageable number. Using *Google Scholar* for the same search, I was able to cut the number of Web results from nearly 20,000 in the original search to 243. As long as you choose your keywords carefully, the most relevant pages will usually be at the top of the list that comes back. *Scirus* is also an excellent engine for conducting keyword searches. *Google, AltaVista*, and *Ixquick* provide easy access to drawings, photographs, and videos: Click on "Images" from the home page.

At most sites, including *Yahoo!*, you can begin with a directory search, and once you get the topic sufficiently focused, you can switch to a keyword search. For example, starting with a directory search and clicking first on Science and then on Biology (or Life Sciences), I could then type the keyword *journals* in the search box and gain access to online journals in the area of interest.

You can also use keyword searches to locate specific papers. Consider this paper, published in 2010 by Tseng, Beane, Lemire, Masi, and Levin: "Induction of vertebrate regeneration by a transient sodium current." Try locating the pdf for this paper using the fewest number of keywords possible. See Technology Tip 2 for hints about choosing a strategy.

Metasearch engines, which explore the databases of many other search engines simultaneously, are generally to be avoided. The databases they include are often inadequate for scientific searches or the listings are sometimes paid for by commercial enterprises. Two exceptions (see Table 1) are *Ixquick* and *SurfWax*.

The Web is growing in size and complexity with such rapidity that I can give only general advice about Web searches here. For more specific information visit any of the sites listed in Appendix C. That listing includes two sites that periodically update and evaluate search engines.

TECHNOLOGY TIP 2

Using search engines effectively

Using Web-based search engines effectively takes practice. If your search terms are too general, you might end up with thousands or even tens of thousands of hits. The trick is to narrow your search, and to do so efficiently.

1. **Don't enter search terms in complete sentences.** Using *Google*, for example, there is no need to ask, "Does bicycle riding cause impotence in males?" Instead, just type the following: *bicycle seats impotence*. If you want a definition for the word *osculation*, just type in the following: *define osculation*.
2. **Use quotation marks to search for specific phrases.** To find information about the invasive zebra mussel, type in the following: *"zebra mussel" invasions*. Without the quotation marks, your results will include many references to zebras.
3. **All search engines allow you to specify particular terms or phrases to include or exclude,** typically using a plus or minus sign or the words *AND* or *NOT*. In some search engines, you would type *parasites + salmon – nematodes* to restrict your search to include parasites of salmon that are not nematodes. In other search engines, you would conduct the same search by typing *parasites AND salmon NOT*

nematodes. Check the help pages within specific search engines for additional information, or consult http://www.searchengineshowdown.com/features/.

4. **Narrow your searches further using Advanced Search.** Once you have a good idea of the type of material you are searching for, you can focus your searches on that material. With both *Google* and *Scirus*, for example, you can limit your search to particular file types (e.g., PDF files, or PowerPoint® files) and particular domains (e.g., .edu files or .gov files). With *Google Scholar* and *Scirus*, Advanced Search allows you to search for articles by particular authors, or with particular words in the text or titles, or that were published in particular years, and to restrict your search to particular subject areas (e.g., Biology, Life Sciences, or Environmental Science).
5. **Broaden your search using wildcards.** Many search engines allow you to use "wildcards" for this purpose. If, for example, you were using the ISI Web of Knowledge and wanted to find all articles that included the words *parasite, parasites, parasitism,* and *parasitic* in their titles, you could enter the following: *paras**. In this case, the asterisk substitutes for any of those word endings. If you wanted to limit your search to nematodes, a particular group of parasites, you could enter the following: *para* AND nematod**. You cannot search this way using *Google*, but *Google* automatically includes word variants in its searches.
6. **If you know the title of a particular paper that you are looking for in electronic format, you can enter all or part of the title in quotes so that the engine doesn't search each word individually.** For example, if I were looking for a particular paper published in 2000 by Michael Barresi and collaborators in the journal *Development*, I might enter just the first part of the paper's title in quotes using *Google* or *Google Scholar*: *"The zebrafish slow-muscle-omitted gene product is required."* Alternatively, you could get to the same published article by including the author's last name in your search and entering something like this: *Barresi "slow-muscle-omitted gene product."* One advantage

(*Continued*)

of searching this way instead of using a formal scholarly database such as the ISI Web of Science is that some papers are available for free through biology department Web sites or individual researcher's Web sites that are not readily available online in full text elsewhere or are available only for a fee.

7. **Use several search engines for each Web search.** Different engines search and index the Web in different ways and cover remarkably different databases.

In addition, *Google, Yahoo!*, and most other search engines provide detailed advice about how best to conduct searches using their databases. You can access this advice from their home pages.

FINAL THOUGHTS ABOUT EFFICIENT SEARCHING: TECHNOLOGY ISN'T EVERYTHING

Cruising the Web is lots of fun, and it's always exciting to use the latest technology. But you can easily spend hours prowling the Internet and return with little of value, especially compared to what you could have gained from spending the same amount of time reading books or articles or even just looking through current issues of relevant journals. Not only is much on the Web not worth reading, most of it is also ephemeral. Web sites appear and disappear, and the information at any particular site can change daily. You can't substantiate any statements of fact or opinion with anything so impermanent. Searching the Web is a great way to find inexpensive airfares and textbooks as well as reviews of the latest movies. It's also an excellent way to learn about particular graduate programs and the research being done by the faculty in those programs. But it's a generally untrustworthy source of information for formal papers and research reports. Unless your instructor says otherwise or asks you to visit specific Web sites relevant to your course, do not cite Web pages as references. It's fine to cite online journals, on the other hand, because the articles in those journals are peer-reviewed before publication.

Remember, **your goal in conducting a literature search is generally to collect a small number of high-quality references that you will then read carefully** so that you can discuss them with conviction. Returning from a voyage on the Internet bearing hundreds or even thousands of references creates only an illusion of accomplishment. Yes, you've located many megabytes of information. But the real work—and the really useful work—comes after you select your references and sit down to read them. Make sure you give yourself plenty of time for that job.

CLOSING THOUGHTS

Electronic databases are good sources of references, but your library will probably not subscribe in print form or for online access to all the journals included in the literature searched by the various services. You may thus spend considerable time accumulating a long list of intriguing references, creating the comforting illusion that you are getting something done, only to discover that you can't access the journals. **Consulting recent issues of available, appropriate research journals (either physically or online) may thus be the most efficient way to find promising research topics and associated references for most undergraduate writing projects.**

SUMMARY

1. Be efficient in exploring the primary scientific literature: Browse the list of references given in your textbook and in other relevant books, and in papers published in recent issues of relevant scientific journals.
2. Become familiar with the major abstracting and indexing services, and use these as necessary to complete your literature search. Be cautious about the validity of information posted on the Internet, and don't be so dazzled by the Internet that you confuse downloading information with reading and understanding it.

3

GENERAL ADVICE ON READING AND NOTE TAKING

The truth is that badly written papers are most often written by people who are not clear in their own minds what they want to say.

JOHN MADDOX

That is one of the worst feelings I can think of, to have had a wonderful moment or insight or vision or phrase, to know you had it, and then to lose it. So now I use index cards.

ANNE LAMOTT

What Lies Ahead? In This Chapter, You Will Learn

- Questions to ask yourself when looking at figures and tables
- How (and why) to summarize information in your own words as you read
- How to take notes effectively, to increase understanding and avoid plagiarism
- The importance of documenting your sources as you read

WHY READ AND WHAT TO READ

As discussed in Chapter 2, the **primary literature** presents original observations and experiments, and includes detailed information about how those observations or experiments were made or conducted. Articles encompassing this primary literature are published (after being reviewed and evaluated by other scientists—a process called peer review) in formal research journals such as *Science, Ecology, Genome Biology*, and *BMC Neuroscience*. The **secondary literature**, in contrast, is based on summaries of the primary literature. Textbooks, magazine articles, encyclopedia entries, and review articles are examples of the secondary literature.

Reading secondary sources is an excellent way to get up to speed in a new area, but you will generally be learning more about what is known than about what remains to be found out. For most assignments in upper-level biology courses, you will be asked to go beyond the factual foundation of the field and to interpret, evaluate, synthesize, ask new questions, and maybe even design experiments to address some of those questions. To see the basis for—and often the limits of—our current knowledge, you will need to explore the primary research literature. Reading that literature is very different from reading the secondary literature. It's a skill that requires some practice and a bit of guidance.

EFFECTIVE READING

Too many students think of reading as the mechanical act of moving the eyes left to right, line by line, to the end of a page, and repeating the process page after page to the end of a chapter or an assignment. I call this "brain-off" reading. When the last page has been "read," the task is over and it's on to something else. This is, after all, the way we typically watch television: We sit transfixed before the television until the program has ended and then either change the channel or turn the set off; we've "seen" the program. In the same way, students typically "listen" to a lecture by furiously copying whatever the instructor writes or says, without really thinking about the information as it is presented.

However, **if you hope to develop something worth saying in your writing, you must interact intellectually with the material**; you must become what I call a "brain-on" reader, wrestling thoughtfully with every sentence, every graph, every illustration, and every table *as you read*. If you don't fully understand some element of what you are reading (including your lecture notes), you must work through the problem until it is resolved rather than skipping over the difficult material and moving along to something more accessible.

This is inevitably a time-consuming process, but you can do a number of things to smooth the way. Whether you are writing an essay or a review paper, the Introduction or Discussion section of a laboratory report or research article, or the introduction to an oral presentation, always begin by carefully reading the appropriate sections of your textbook and class notes to get a solid overview of your general subject. It is usually wise to then consult 1 or 2 additional textbooks or review articles before venturing into the primary literature to read about the results of original research; a solid construction requires a firm foundation. Your instructor may have placed a number of pertinent textbooks on reserve in your college library. Alternatively, you can consult the library online catalog, looking for books listed under the topic you are investigating. Science encyclopedias, such as the *McGraw-Hill Encyclopedia of Science and Technology*, are also excellent sources of factual information. Wikipedia articles can be good sources of background information, but as mentioned earlier (p. 27), you should always be skeptical of their accuracy because you don't know the source of the information.

Armed with this background information, you are now prepared to tackle the primary research literature. See Chapter 2 for advice on locating this literature.

Reading a formal scientific paper is unlike reading a work of fiction or even a textbook or review article. The primary scientific literature must be read slowly, thoughtfully, and patiently, and a single paper must usually be reread several times before it can be thoroughly understood. Don't become discouraged after only 1 or 2 readings. As with playing tennis or sight-reading music, reading the primary literature gets easier with practice. If, after several rereadings of the paper, and if, after consulting several textbooks, you are still baffled by something in the paper you are reading, ask your instructor for help.

As you carefully read each paper, pay special attention to the following:

- What specific questions were asked?
- How was the study designed, and how does the design of the study address the questions posed?
- What are the controls for each experiment?
- What are the specific results of the study? How convincing are they? Are any of the results surprising?
- What assumptions were made? Do they seem reasonable?
- What contribution does the study make toward answering the original question?
- What aspects of the original question remain unanswered?

You can answer many of these questions just by studying the figures and tables in the Results section, as described below.

If you are planning to write a research proposal (see Chapter 10), you should ask one additional question:

- What specific question(s) would I ask next?

As discussed below, you should spend most of your time looking at the original data presented in the papers you read.

READING DATA: PLUMBING THE DEPTHS OF FIGURES AND TABLES

Data—the most important part of any book or journal article—are displayed either in figures or in tables; both must be examined critically. Your goal here is to come to your own interpretation of the data so that you can better understand or evaluate the author's interpretation. To do this, **you must study the data and ask yourself some questions about how the study was done, why it was done, and what the major findings were**.

Consider the example shown in Figure 1, modified from a review paper by David E. Cochrane entitled "Peptide regulation of mast-cell function." From your background reading or class lecture notes, you would probably know that mast cells release into the blood a variety of substances involved in provoking allergic and inflammatory responses. If you didn't already know this, you would do some background reading in your textbook before proceeding.

Looking at Figure 1, let's see if we can figure out what the researchers did to collect their data. By reading the axis labels, we learn that the graph shows how blood histamine concentrations changed over time, and we learn from the figure caption that these changes were provoked by a particular peptide called neurotensin. The study was done on anesthetized rats, and all the action seems to have occurred rather quickly, since the *x*-axis extends only to 30 minutes. Looking more closely, we see that histamine concentrations were initially quite low (less than 1 nanogram* per 10 microliters** of blood plasma), as indicated by the arrow (at the lower left side of the graph), and that they rose impressively by the time the first blood sample was taken 1 minute into the study. Other blood samples were taken 3, 15, and 30 minutes after neurotensin was administered, and 3 to 8 separate samples were taken at each time period. Even without reading the figure caption and without seeing the numbers alongside each data point, we would know that replicate samples were taken, because the thin lines extending vertically from each point indicate the amount of variation seen about the mean value recorded for each sampling time; obviously, one can see variation about a mean value only when multiple samples were taken.

What were the controls for this experiment? Apparently, a number of rats were injected with a saline solution instead of with neurotensin, and blood samples were also taken from these rats at appropriate times.

So, we know quite a bit about how this aspect of the study was conducted just by scrutinizing the graph. It helps considerably that the graph and figure caption were carefully constructed. What additional

*1 nanogram (ng) = 10^{-9} gram.
**1 microliter (μl) = 10^{-6} liter.

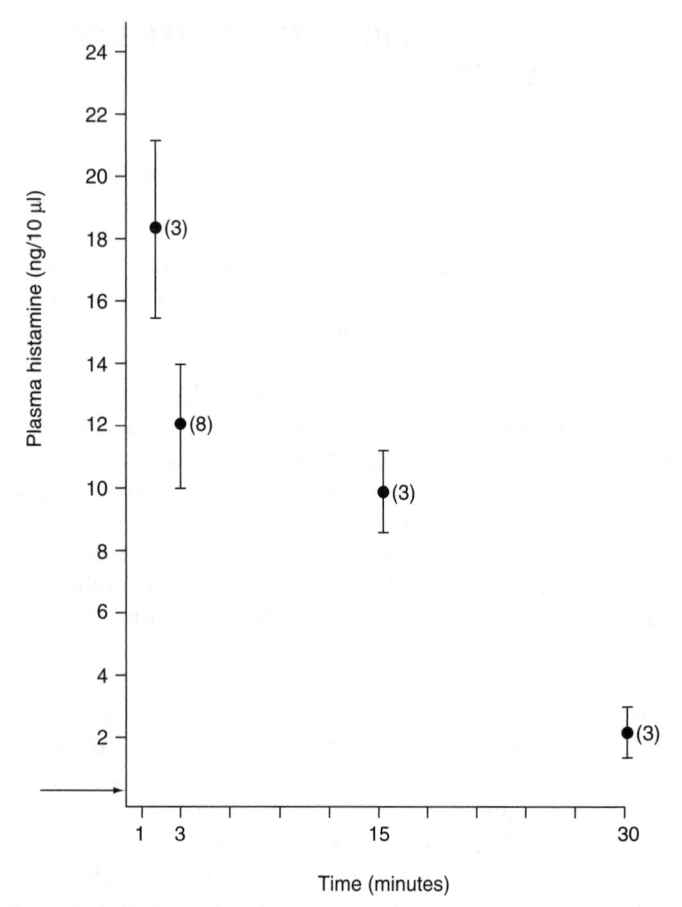

Figure 1. Plasma histamine concentrations in response to neurotensin (NT) given at t = 0. Rats were anesthetized and given NT (5 nmol/kg) or saline (0.3 ml) intravenously. Blood samples were collected at the indicated times. Each point represents the mean (one standard error about the mean) of n values (given in parentheses). The horizontal arrow (lower left) shows the mean histamine concentration before addition of NT. Intravenous injection of saline did not alter this concentration over the 30-minute observation period.

Source: From Cochrane, D.E., 1990. Peptide regulation of mast-cell function. In *Progr. Medicinal Chemistry*, Vol. 27; G.P. Ellis and G.B. West, eds.; Elsevier Science Publ. (Biomedical Division), pp. 143–188.

information might you wish to have? Here are a few questions that might arise as you think about the figure:

1. How did Dr. Cochrane decide to inject neurotensin at a concentration of 5 nmol/kg? Is this a physiologically realistic concentration?
2. How was histamine concentration determined?
3. How many rats served as controls for each time interval?

4. How old were the rats? Which sex were they? Where did Dr. Cochrane get them?
5. How were the rats anesthetized, and might this pretreatment have affected their response to neurotensin?
6. With regard to the replicate samples, were 3 to 8 blood samples taken simultaneously from a single rat at each sampling time, or were blood samples taken from 3 to 8 different rats each time? Blood samples were *probably* taken from a number of different individual rats, but we can't be sure from the graph.
7. Were different rats sampled at each time interval, or were the same individuals sampled repeatedly—for example, at 15 minutes and again at 30 minutes? One might guess that the same individuals were sampled repeatedly. On the other hand, there may be technical limitations to drawing blood from any individual rat more than once. Suppose, for instance, that a very large volume of blood was required to determine the histamine concentration. This issue leads to my last 3 questions.
8. What volume of blood was withdrawn for each sample?
9. How were the blood samples obtained? How was the plasma separated from the blood cells before analysis?
10. Does drawing blood influence the production of histamine? How could one tell?

Asking such questions puts you in a good position to interpret the results illustrated and provides a framework for your later reading of the Materials and Methods section of the report; you can now enter the methodology section of the paper looking for the answers to specific questions.

If asked to describe how this aspect of the study was conducted and why the study was probably undertaken, how might you respond? Even without reading anything else in the paper, **you could write the following summary**:

> This study was apparently undertaken to determine the ability of the peptide neurotensin to elicit histamine secretion from mast cells. A number of rats were anesthetized and then injected intravenously with 5 nmol/kg of neurotensin, whereas others (control rats) received intravenous injections of saline solution to control for the possibility that the act of injection itself provoked histamine release. Some quantity of blood was withdrawn from a number of rats 1, 3, 15, and 30 minutes after injections were made, and the histamine concentration in each blood sample was somehow determined.

Now, in a few short sentences, let us try to summarize the results, beginning with the most general statements we can make.

1. Neurotensin had a dramatic and rapid effect on histamine concentrations in the blood of laboratory rats, with histamine

concentration increasing by about 20 times within 1 minute after injection.

2. The effect seems rather short-lived; only 3 minutes into the study, the histamine concentrations had already fallen considerably from the peak level recorded 2 minutes earlier.
3. By 30 minutes after the injection, the mean histamine concentration approached initial levels.
4. The effect shown is clearly due to the peptide rather than the procedure itself, because histamine concentrations in the blood of control rats receiving saline injections did not change appreciably during the study.

Summarizing information as you read is one of the best ways to digest the material and generate new ideas, as discussed in the next section.

You can approach tabular data in the same way, first reading the accompanying legend and column headings and then asking yourself *how* the study was done, *why* it might have been undertaken, and *what* the key general results seem to be. If the tables were prepared with as much care as the figure we just looked at, you should be able to answer each of those questions.

For example, take a look at Table 1. For some reason, the authors wanted to know if 3 types of predatory cats in Peru ate the same foods, and they came up with an ingenious way of addressing that question. I'll let you think about why they might have asked the question.

Reading the table legend, you know where the study was conducted, what the data represent, and what "n" stands for. Looking at the table, you know what predators were of interest and the 7 categories of prey found in their feces. How do you think the study was done?

Looking now at the data, how much did the predators overlap in diet? Did each predator specialize on any one type of prey? Did all 3 predators specialize on the same prey? Did each predator eat at least some prey from each category? The more you look at the data, the more you will see, and the more questions you will ask. What questions would you like to ask the authors? What question would you ask next if you were going to continue this study?

Apply the same procedure figure by figure, table by table, until you reach the last bit of data. Now it is safe to actually read the text of the paper. Perhaps you will find the author reaching conclusions different from your own. You may have missed some crucial element in studying the data yourself, or perhaps you will learn something crucial in the text of the paper that was not made clear in the figure. If so, you can then happily leave your own opinion behind and embrace the author's. On the other hand, you may genuinely disagree with the conclusion reached by the paper's author (or authors), or you may object to the author's interpretation because of some

Table 1. The diets of 3 predatory cats from Rio Manu, Peru, inferred from fecal samples collected in nature. An entry of 5% (top left) means that fish, in this case, made up 5% of all the prey items found in the feces from that predator. The data are based on the minimum number of prey that must have been eaten to obtain the number of bones of each type collected. n = number of feces examined from each kind of predator.

	Name of predator			
Kind of prey	Jaguar (n = 40)	Puma (n = 12)	Ocelot (n = 177)	Total eaten from all predators
Fish (%)	5	0	2	7
Reptiles (%)	33	17	12	62
Small rodents and opossums (%)	8	17	66	91
Large rodents (%)	15	58	5	78
Other large mammals (%)	23	0	0	23
Birds (%)	10	0	11	21
Bats (%)	5	8	5	18
Total (%)	99	100	101	

Based on Emmons, L.H. 1987. *Behav. Ecol. Sociobiol. 20: 271–283.*

concern you have about the method used. This is, in fact, how new questions often get asked in science, and how new studies get designed, with each new step building on the work of others. Most of us will never have the final say in any particular research area, but we can each contribute a valuable next step, even if our individual interpretation of that step turns out to be mildly—or even dramatically—wrong.

If you think about figures and tables in this way, you may be on your way to a productive scientific career, research project in hand. At the very least, you will be in an excellent position to *discuss* the paper, either on its own or in relationship to other studies that you will go on to scrutinize just as thoroughly.

If you don't go through these steps in "reading" the data, you will be all too accepting of the author's interpretation. In consequence, you will have considerable difficulty avoiding the book report format in your writing, simply repeating what others did and what they say they found. You can move your work to a higher, more interesting plane (more interesting for you and for the reader) by becoming a brain-on reader. It's tough slogging, and although it becomes considerably easier with practice, it never becomes trivial. But it does become enjoyable, and satisfying, in the same way that a good tennis game can be enjoyable and satisfying even when played in hot, humid weather.

READING TEXT: SUMMARIZE AS YOU GO

Resist the temptation to copy your source's words verbatim or simply to highlight them. Instead, **try to summarize chunks of material as you read along,** as in the neurotensin example discussed earlier (pp. 39–40). In that way, you will be processing information as you read it, which moves you one major step closer to having something interesting to write about later.

To summarize effectively, you must first determine the most important points in the material you wish to summarize. Consider the following paragraph from Rachel Carson's landmark book *Silent Spring.** For many people, this book, first published in 1962, marks the start of the environmental movement.

> Water, soil, and the earth's green mantle of plants make up the world that supports the animal life of the earth. Although modern man seldom remembers the fact, he could not exist without the plants that harness the sun's energy and manufacture the basic foodstuffs he depends upon for life. Our attitude toward plants is a singularly narrow one. If we see any immediate utility in a plant we foster it. If for any reason we find its presence undesirable or merely a matter of indifference, we may condemn it to destruction forthwith. Besides the various plants that are poisonous to man or his livestock, or crowd out food plants, many are marked for destruction merely because, according to our narrow view, they happen to be in the wrong place at the wrong time. Many others are destroyed merely because they happen to be associates of the unwanted plants.

What are the key points in this paragraph? What information would Carson be unhappy to see left out?

- All animals, including people, depend on plants for food.
- People think little about the consequences of destroying plants that annoy them or don't seem to do anything useful.

Here is a possible one-sentence summary that incorporates both of these points:

> People destroy any plant that happens to annoy them or doesn't seem to do anything useful, forgetting the extent to which all animal life ultimately depends on photosynthesis.

---

*I have not altered the original wording, with its anachronistic use of what is now generally considered to be sexist writing. If Rachel Carson were writing this today, she would most likely replace "modern man" with "people," and "he" with "we." This would actually strengthen the paragraph, by pointing the finger clearly at all of us.

This summary is (1) **accurate**, (2) **complete**—it incorporates all major points, (3) **self-sufficient**—it makes good sense even if the reader has never read the original text, and (4) **in my own words**. Get in the habit of writing such summaries as you read and as you take notes during lectures. It is a real challenge, but one that gets easier with practice. Persist: The eventual payoff is tremendous. Moreover, summarizing in your own words is an excellent way to avoid plagiarism, as discussed below (see also pp. 13–15).

TAKE NOTES IN YOUR OWN WORDS

Photocopying an article or book chapter does not constitute note taking; neither does highlighting or even copying a passage by hand, occasionally substituting a synonym for a word used by the source's author. Take notes using your own words; you must get away from being awed by other people's words and move toward building confidence in your own thoughts and phrasings.

Note taking involves critical evaluation; as you read, you must decide either that particular facts or ideas are relevant to your topic or that they are irrelevant. As Sylvan Barnet says in *A Short Guide to Writing about Art* (2003, Pearson Longman, 7th edition, p. 251), "You are not doing stenography; rather, you are assimilating knowledge and you are thinking, and so for the most part your source should be digested rather than engorged whole." If an idea is relevant, you should **jot down a summary using your own words** (see p. 42). Try not to write complete sentences as you take notes; this will help you avoid unintentional plagiarism later and will encourage you to see through to the essence of a statement while note taking. For the same reason, **do not take notes or write while you are looking at the source**.

Sometimes an author's words seem so perfect that you cannot see how they might be revised to best advantage for your paper. In this case, you may wish to copy a phrase or a sentence or two verbatim, but be sure to enclose this material in quotation marks as you write, and clearly indicate the source and page number from which the quotation comes. If you modify the original wording slightly as you take notes, you should indicate this as well, perhaps by using modified quotation marks: ″/. . . ″/. If your notes on a particular passage are in your own words, you should also indicate this as you write. I precede such notes, reflecting my own ideas or my own choice of words, with the word *Me* and a colon. If you take notes in this manner, you will avoid the unintentional plagiarism that occurs when you later forget who is actually responsible for the wording of your notes or for the origin of an idea.

If you find yourself copying verbatim or paraphrasing your source, be sure it is not simply because you do not understand what you are reading.

Be honest with yourself. **It is always best to summarize in your own words as you read along**; at the very least, you should think your way to some good questions about what you are reading, and then write those questions down. Sooner or later, serious intellectual engagement is required; there are no shortcuts available here, I'm afraid.

Here is an example of some notes taken using the suggested system of notation. These notes are based on a paper by William Biggers and Hans Laufer. Figure 2 shows an excerpt from the paper, and Figure 3 shows some notes based on that excerpt. Note that **the student has avoided using complete sentences**, focusing instead on distilling the basic points and pinning down a few words and phrases that might be useful later. Notice also that **the student has taken notes selectively**, has found it unnecessary to quote any of the material directly, and **has clearly distinguished his or her own thoughts** from those of the original authors (by preceding such thoughts with *Me:*). Similarly, the student has distinguished between what the authors did and what the student thinks could be done later or might have influenced the results. Perhaps the most important points are that the note taker clearly *thought* while reading and took notes in his or her own words. This student will not have to worry about accidental plagiarism when writing a paper based on these notes.

You cannot take notes selectively or in your own words if you do not understand what you are reading. I suggest that you first consult at least one general reference textbook and read that material carefully, as recommended earlier. Once you have located a particularly promising scientific article, read the entire paper at least once without taking any notes. **Resist the (strong) temptation to annotate and take notes during this first reading**, even though you may think that without a pen in your hand, you are accomplishing nothing. Put your pencils, pens, note cards, paper, or laptop computer away, and just read. **Read slowly and with care.** Read to understand. Study the illustrations, figure captions, tables, and graphs carefully, and try to develop your own interpretations before reading those of the author(s). Don't be frustrated by not understanding the paper at the first reading; understanding scientific literature takes time and patience—and often many rereadings, even for practicing biologists. Concentrate not only on the results reported in the paper but also on the reason the study was undertaken and the way the data were obtained. **The results of a study are real; the interpretation of the results is always open to question.** And the interpretation is largely influenced by the way the study was conducted. Read with a critical, questioning eye. Many of the interpretations and conclusions in today's journals will be modified in the future. It is difficult, if not impossible, to have the last word in biology; progress is made by continually building on and modifying the work of others.

Introduction

The chemoreception by marine invertebrate larvae of chemical "cues" that are present in the ocean environment and induce settlement and metamorphosis is important for the recognition of habitats that favor growth and reproduction (Chia and Rice, 1978; Rittschof and Bonaventura, 1986; Scheuer, 1990). [1] These settlement signals appear to be specific for different species, as evidenced by findings that larvae of the abalone *Haliotis rufescens* respond to specific chemicals in red algae (Morse *et al.,* 1984), larvae of the nudibranch *Phestilla sibogae* respond to chemicals in corals (Hadfield, 1978, 1984), larvae of the polychaete annelid *Phragmatopoma californica* respond to chemicals present in the burrows of adult worms (Pawlik, 1988, 1990; Jensen and Morse, 1990), and larvae of the sand dollars *Dendraster excentricus* (Burke, 1984) and *Echinarachnius parma* (Pearce and Scheibling, 1990) respond to chemicals produced by adult sand dollars. [2]

In previous studies, we have found that juvenile hormones (JH), which are known morphogens that regulate reproduction and development of insects and crustaceans (Laufer and Borst, 1983, 1988; Laufer *et al.,* 1987), as well as chemicals with juvenile hormone activity in insect cuticle bioassays, are able to induce settlement and metamorphosis of metatrochophore larvae of the polychaete annelid *Capitella* sp. I (Biggers and Laufer, 1992, 1996). In nature, larvae of *Capitella* sp. I are stimulated to settle and [3] metamorphose when they come into contact with chemical inducers present in sediments (Butman *et al.,* 1988); although the identity of these chemicals remains in debate (Cuomo, 1985; Dubilier, 1987), they appear to have JH-activity (Biggers, 1994).

We have now investigated the signal transduction process through which the *Capitella* larvae respond to JH-active compounds. Our results presented in this paper indicate that JH-active compounds stimulate settlement and metamorphosis of these larvae through the activation of protein kinase C (PKC) and subsequent [4] modulation of ion channels.

Materials and Methods

Capitella *larval settlement bioassays*

Stock cultures of *Capitella* sp. I were maintained at 18°C in artificial seawater (Utikem Co.) and washed sea sand (Fisher Scientific) and were fed Tetramin fish food flakes. Brood tubes containing adult females along with their developing eggs and larvae were then separated from the cultures and placed into 60-mm glass Petri dishes containing seawater. The dishes were checked daily for hatched, swimming larvae to be used for bioassays. Stock solutions of juvenile hormone III (JH III), MF, phorbol-12,13 dibutyrate (PDBU), 1-(5-isoquinolinyl-sulfonyl)-2-methylpiperazine (H-7), arachidonic acid, elaidic acid, verapamil, 4-aminopyridine, and nigericin were prepared in 95% ethanol. Settlement and metamorphosis bioassays were conducted at 18°C using 60-mm glass Petri dishes that were pre-baked at 250°C to remove contaminants (Biggers and Laufer, 1996). For the assays, 10 to 100 μl of the test chemical stock solutions were added by micropipette into Petri dishes containing 10 metatrochophore larvae less than 1 day old (1 day post-release), and 10 ml of artificial seawater. The dishes were then observed. After 1 h, the amount of settlement and metamorphosis was as- [5] sessed by placing each dish under a dissecting microscope and counting the number of settled larvae crawling on the bottom of the dish. Metamorphosis after 1 hour was also more critically assessed by using a compound microscope and noting the loss of cilia, elongation, and development of capillary setae.

Protein kinase C assays

Assays for the presence of protein kinase C were carried out essentially as described by Yasuda *et al.* (1990), by measuring phosphorylation of an 11-residue synthetic peptide from myelin basic protein (MBP_{4-14}). This method is specific for measurement of PKC, and permits selective measurement in crude tissue prepar-

Figure 2. Modified excerpt from "Settlement and metamorphosis of *Capitella* larvae induced by juvenile hormone-active compounds is mediated by protein kinase C and ion channels," by William J. Biggers and Hans Laufer (pp. 187–188), as reprinted from *The Biological Bulletin* 196 (1999): 187–198. Reprinted with permission of *The Biological Bulletin* and the authors, William J. Biggers and Hans Laufer.

1. Marine larvae do not metamorphose until contact chemicals that indicate juveniles will do well here.

2. Larvae of different species respond to different chemicals.
   Me: How many diff. species have been looked at?
   Do we know what the chemical cues are? How (and where?) do the larvae sense them?

3. Larvae of the polychaete Capitella sp. I metamorphose in response to a juvenile hormone-like chemical in mud.
   Me: odd that same chemical works on both insects and worms?
   What distinguishes polych. from other annelids?
   And how big are these larvae? Microscopic I think...

Purpose of this study: How does the juvenile hormone make the larvae metamorphose?

4. Apparently the chem. cue activates "protein kinase C," which then either opens or closes ion channels somewhere (Me: calcium channels maybe?).

5. Methods: Collect larvae → expose to very small amounts of chemical (≤ 100 µl*!) $\xrightarrow{1\,hr}$ count number of larvae that have metam.'d. Nine chemicals tested (Me: must find out what each one does).

Figure 3. Handwritten notes based on the article by Biggers and Laufer (see Figure 2). Numbers in the margin correspond to the indicated portions of Figure 2.

Once you have completed your first reading of the paper, you may find that the article is not really relevant to your topic after all or is of little help in developing your theme. If so, the preliminary read-through will have saved you from wasted note taking.

People differ in their note-taking styles. Some people suggest taking notes on index cards, with one idea per card so that the notes can be sorted readily into categories at a later stage of the paper's development. If you prefer taking notes on full-size paper, begin a separate page for each new source, and write on only one side of each page to facilitate sorting into categories later. Similarly, if you enter notes directly into a computer, be sure to leave a few lines of space above and below each entry. Whatever system works best for you, that's the one to use. If your present system of note taking is not working, experiment until you find one that does.

---

*microliters

Split-Page Note Taking: A Can't-Fail System

> *Perfectly organized notes that cover everything are beautiful, but they live on paper, not in your mind.*
>
> PETER ELBOW

If you have trouble taking notes in your own words and thinking as you read, try split-page note taking. With this system, you divide a piece of paper into left and right halves, either by folding the paper in half lengthwise or by drawing a line down the middle. On the left side of the page, write factual information as you read—preferably in your own words, but it's also okay to quote directly. Then, on the right side of the page, write your response to the entry you just made on the left side. Try to respond to everything that you write on the left as you read. Your response could be a simple question ("What on earth is protein kinase C?"), a more thoughtful question ("How can they know that? First they would have had to measure … "), a reminiscence ("That reminds me of what Professor Bolker said in lecture last week about … "), or a comparison ("Interesting: The freshwater species have very different life histories"). Write whatever you happen to think of when you look at what you wrote on the left. See Figure 4 for an example of split-page note taking in action, based on the first few paragraphs of the material presented in Figure 2.

Split-page note taking may seem like a gimmick. Well, actually it *is* a gimmick, but it's a gimmick that *works*. It works by slowing you down, forcing you to think as you read. Split-page note taking is also a very effective way to read tables and figures. Try it, for example, in looking at Figure 1 and its caption. Even experienced note takers often find that they learn something new by taking notes this way.

FINAL THOUGHTS ON NOTE TAKING: DOCUMENT YOUR SOURCES

As you take notes, be sure to make a complete record of each source used: author(s), year of publication, volume and page numbers (if consulting a scientific journal), title of article or book, publisher, and total number of pages (if consulting a book). It is not always easy to relocate a source once it has been returned to the library stacks; in fact, the source you forgot to record completely is always the one that vanishes as soon as you realize that you need it again. Finally, before you finish with a source, it is good practice to read through that source one last time to be sure that your notes accurately reflect the content of what you have read.

Facts	Responses
marine invert. larvae are induced to metam. by chemical cues	True for all larvae? How many species have been studied? The larvae are awfully tiny—what body parts "smell" the cues? What are the cues?
Different species respond to diff. chem. cues	Must be pretty widespread. Examples given incl. a variety of species from very different groups.
Juvenile hormone (JH) regulates insect development, but also makes polychaete annelid larvae metamorphose	Why would juvenile hormone be floating around in seawater? Maybe it's not—maybe it's not the chem. cue discussed in first paragr. but rather something that acts inside the larva? Why would an insect chemical work on a worm? Are worms and insects closely related?

Figure 4. An example of split-page note taking, based on the first 2 paragraphs of material presented in Figure 2. The student has recorded factual information on the left side of the page and her response to that information on the right.

SUMMARY

1. Become a brain-on reader: Work to understand your sources fully, sentence by sentence, figure by figure, and table by table.
2. Take notes thoughtfully. In particular, practice summarizing information as you go along. Your summary must be accurate, complete, self-sufficient, and in your own words.
3. In your note taking, be careful to distinguish your words and thoughts from those of the author(s) to avoid unintentional plagiarism. Be sure to record the complete citation information for everything on which you take notes.

4

READING AND WRITING ABOUT STATISTICAL ANALYSES

It is difficult to read the primary biological literature without running into statistics. And it is virtually impossible to draw conclusions from most laboratory or field studies involving numerical data without subjecting those data to statistical analysis. This chapter is no substitute for a full course in biostatistics, but it will get you off to a good start. Appendix C lists some additional references on this topic.

What Lies Ahead? In This Chapter, You Will Learn

- What is meant by the terms *statistical analysis* and *statistical significance*
- How biologists use statistics to deal with variability
- What a null hypothesis is, and how it can be evaluated
- That statistical analysis cannot prove things; it can only support the null hypothesis or call it into question
- How to write about statistics without drawing attention away from the biology involved in the study

STATISTICAL ESSENTIALS

Variability and Its Representation

Variability is a fact of biological life: Student performance on any particular examination varies among individuals; the growth rate of tomato plants varies among seedlings and from place to place and year to year, or even week to week; the effects of a particular concentration of a particular pollutant vary among species, and among individuals within a species; the respiration rate of mice held under a given set of environmental conditions varies among individuals; the number of snails occupying a square meter of substrate varies from place to place and from year to year; the extent to which a particular chemical enhances or suppresses the transcription of a particular gene varies from test tube to test tube; and the amount of time a lion spends feeding varies from day to day and from lion to lion. Of course, some of the variability we inevitably see in our data reflects unavoidable imprecision in making measurements. If you measure the length of a single bone 25 times to the nearest millimeter (mm), for example, you will probably not end up with 25 identical measurements. But much of the variability recorded in studies reflects real biological differences among the individuals in the sample population. Put identical meals in front of 20 people in a restaurant, and few of these people will finish their meals at the same time. Moreover, the amount of food consumed will probably also vary quite a lot among individuals. This sort of natural variability is referred to as "error" by statisticians, but it isn't "error" in the sense of "making mistakes." It's better to think of such variation as natural "scatter" in the data (Motulsky, 1995; see Appendix C).

Variability, whether in the responses you measure in an experiment or in the distribution of individuals in the field, is no cause for embarrassment or dismay, but it cannot be ignored in presenting or interpreting results.

Suppose you have 2 samples of 4 rats each. The lengths of the rats' tails in samples *A* and *B* are

A = 7.2, 7.0, 6.8, 7.0 cm

B = 3.6, 12.5, 3.3, 8.6 cm

Both samples have the same mean value (7.0 cm), but the tails in sample *A* are much less variable in length than those in sample *B*. Simply listing the mean value, then, would omit an important component of the story contained in the data.

Listing the mean and the range of values obtained in each sample would help, but presenting the **variance** (σ^2) about the mean would give an even better indication of how much the data vary within the populations sampled. Variance is readily calculated using a statistical calculator or computer, but for small sample sizes, the variance is not difficult to calculate by hand. Here is the formula:

$$\sigma^2 = \frac{\sum_{i=1}^{N}(X_i - \overline{X})^2}{N-1}$$

where N is the number of observations made, X_i is the value of the ith observation, $\overline{X}$ is the mean value of all the observations made in a sample, and Σ is the symbol for summation. To calculate the variance, then, you simply sum the squared differences of each individual measurement from the mean of all the measurements and then divide that number by $N - 1$. As an example, suppose you have the following data points:

5 cm

4

4 $\qquad\qquad N = 5$

6

5

$$\overline{X} = \frac{\sum_{i=1}^{N}}{N} = \frac{24}{5} = 4.8 \text{ cm}$$

$$\sigma^2 = \frac{(5-4.8)^2+(4-4.8)^2+(4-4.8)^2+(6-4.8)^2+(5-4.8)^2}{4}$$

$$= 0.7$$

Try making the same calculation for the lengths of the rats' tails in samples *A* and *B* (p. 50).

All we are doing is seeing how far each observation is from the mean value obtained and then adding all those variations together. The squaring is done simply to eliminate minus signs so that we have only positive numbers to work with. Clearly, 100 measurements should give a more accurate estimation of the true mean tail length than only 10 measurements would, and if we had the time and the patience, 1,000 measurements would be better still. We thus divide the sum of the individual variations by a factor related to the number of observations made. Increasing the sample size will reduce the extent of experimental uncertainty. In a sense, then, **variance indicates the amount of confidence we can have in our estimate of the mean**. The smallest possible variance is zero (all measurements were identical); there is no upper limit to the potential size of the variance.

People do not generally report variance per se. Rather, they typically report something related to the variance—the standard deviation, the standard error, or the "95% confidence interval."

To calculate the **standard deviation** (SD), simply take the square root of the variance.

To calculate the **standard error of the mean** (SEM), simply divide the standard deviation by the square root of *N*.

Unlike standard deviation and standard errors, the related **95% confidence interval** has inherent meaning. If you were to repeatedly sample a population 100 times and calculate a mean result for each sample, you can expect 95 of those means to fall within the calculated confidence interval. For sample sizes larger than about 15, the 95% confidence interval is approximately twice the standard error of the mean.

When Is a Difference a Meaningful Difference? What You Need to Know about Tomatoes, Peas, and Random Events?

Suppose we plant 2 groups of 30 tomato seeds on day 0 of an experiment, and the individuals in group *A* ($N = 30$) receive distilled water whereas those in group *B* ($N=30$) receive distilled water plus a nutrient supplement. Both groups of seedlings are held at the same temperature, are given the same volume of water daily, and receive 12 hours of light and 12 hours of darkness (12L:12D) each day for 10 days. Twenty-six of the seeds sprout under the group *A* treatment, and 23 of the seeds sprout under the group *B* treatment. At the end of 10 days, the height of each seedling is measured to the nearest 0.1 cm, and the data are recorded on the data sheet as shown in Figure 5. **Note that the units** (cm; sample size) **are clearly indicated on the data sheet**, as is the nature of the measurements being recorded (height after 10 days). The number of samples taken, or of measurements made, is always represented by the symbol *N*.

The question now is this: Did the mineral supplement affect the seedlings' growth rate? That is, did it make a difference in the height of seedlings by the 10th day after planting?

If all the group *A* individuals had been 2.0 cm tall and all the group *B* individuals had been 2.4 cm tall, we would have readily concluded that

Group *A* seedlings: water only (height, in cm, after 10 days)
2.1 cm, 2.1, 2.0, 2.8, 2.7, 2.4, 2.3, 2.6, 2.6, 2.5, 2.1, 2.8, 2.0, 1.9, 2.8, 2.0, 2.3, 2.2, 2.6, 1.8, 2.0, 2.2, 2.5, 2.4, 2.3, 2.1
Average = 2.3 cm; N = 26 measurements

Group *B* seedlings: water plus nutrients (height, in cm, after 10 days)
2.6 cm, 2.1, 2.0, 2.4, 2.8, 2.6, 2.2, 2.7, 2.4, 2.4, 2.3, 2.2, 2.4, 2.6, 2.4, 2.8, 2.6, 2.5, 2.6, 2.4, 2.6, 2.3, 2.4
Average = 2.4 cm; N = 23 measurements

Figure 5. Data sheet with measurements recorded. Note that units of measurement are indicated clearly.

growth rates were increased by adding nutrients to the water. If each group *A* individual had been 2.3 cm tall and each group *B* individual had been 2.4 cm tall, we might again have suggested that the nutrient supplement improved the growth rate of the seedlings. In the present case, however, there was considerable variability in the heights of the seedlings in each of the 2 treatments, and **the difference in the average heights of the 2 populations was not large with respect to the amount of variation found within each treatment**. The heights of group *A* seedlings differed by as much as 1.0 cm (2.8 – 1.8), and the heights of group *B* seedlings differed by as much as 0.8 cm (2.8 – 2.0), whereas the average height differences between the 2 groups of seedlings was only 0.1 cm (2.4 – 2.3).

The average height of the seedlings in the 2 populations is certainly different. But does that difference of 0.1 cm in average height reflect a real, biological effect of the nutrient supplement, or have we simply not planted enough seeds to be able to see past the variability inherent in the individual growth rates? If we had planted only one seed in each group, the 2 seedlings might have both ended up at 2.6 cm; some seedlings reached this height in both treatment groups, as seen in Figure 5. On the other hand, the seed planted in group *A* might have been the seed that grew to 2.8 cm, and the seed planted in group *B* might have been one of the seeds that grew to only 2.2 cm. Or it might have turned out the other way around, with the taller seedling appearing in group *B*. Clearly, a sample size of one individual in each treatment would have been inadequate to conclusively evaluate our hypothesis. Perhaps 30 seeds per sample would also have been inadequate. If we had planted 1,000 seeds, or 10,000 seeds, in each group, the differences between the 2 treatments might have been even less than 0.1 cm—or the differences might have been substantially greater than 0.1 cm. If only we had planted more seeds, we might have more confidence in our results. If only we had measured 100,000 individuals, or 1,000,000 individuals, or . . .

Wishful thinking has little place in biology, however. We have only the data before us, and those data must be considered as they stand. Is the difference between an average height of 2.3 cm for the group *A* seedlings and 2.4 cm for the group *B* seedlings a real difference? That is, is the difference statistically significant? Or have we simply conducted too little sampling to see through the variability in individual results?

Here is another way to look at the problem. Suppose you stand at the doorway to the Biology or Science building on campus, and you measure the heights of the first 10 men and the first 10 women to enter the building after you arrive at your post. You want to determine whether the men and women on your campus differ, on average, in height. You calculate the mean height for each of the 2 groups, and you find that the means differ. But then you measure the average height of the next 10 men and the next 10 women who enter the building, and you find that the average height of the first and second groups of men also differs, as do the average heights of the first and second groups of women. How can you trust the measured difference in height between men and women when you get different averages from one group of men to the next and from one group of women to the next? That's why we need statistics. **We need statistics whenever we are subsampling from a population in which individuals vary naturally in the traits that we are measuring or whenever individual measurements**—of size, reaction rates, survival, growth, amounts of transcription or translation ... anything that can be measured—**vary from replicate to replicate**. Are the differences between groups real, or do they just reflect how we have unevenly sampled the variation within the populations? Statistics now infiltrate nearly every area of biology.

As an additional example, suppose we have crossed plants producing yellow peas with other plants also producing yellow peas and, from knowledge of the parentage of these 2 groups of pea plants, we expect their offspring to produce yellow and green peas in the ratio of 3:1. Now suppose we actually count 144 offspring that produce yellow peas and 45 offspring that produce green peas, so that slightly more than 3 times as many of the offspring produce yellow peas. Do we conclude that our expectations have been met or that they have not been met? Is a ratio of 3.2:1 close enough to our expected ratio of 3:1? Is the result (144 yellow-producing plants, 45 green-producing plants) statistically equivalent to the expected ratio?

As one final example, suppose we wish to know whether there is a pronounced relationship (a "correlation") between the weight of hermit crabs of a particular species and the size of the shells they occupy in the field. We carefully remove 12 hermit crabs from their shells, weigh the crabs, and then measure the shells in which they had been living. The data are shown in Figure 6. Is there a relationship between hermit crab weight and shell size? How confident can we be in predicting shell size if we know only a hermit crab's weight?

Establishing a Null Hypothesis

Biologists use statistical tests to determine the significance of differences between sampled populations or between the results expected and those obtained. To begin, we typically define a specific issue (hypothesis) to be tested. **The hypothesis to be tested is called the null hypothesis, H_0. The null hypothesis usually assumes** that nothing unusual will

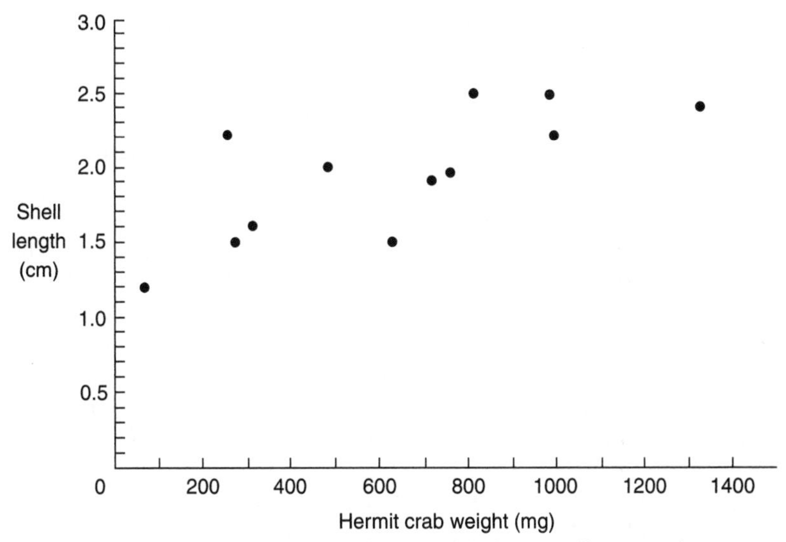

Figure 6. Relationship between wet weight (mg) of the hermit crab *Pagurus longicarpus* and the size of the periwinkle shells (*Littorina littorea*) it occupied at Nahant, Massachusetts, on September 23, 2002 ($N = 12$).

happen in the experiment or study; that is, it assumes **that the treatment—addition of nutrients, for example—will have no effect**, or that there will be no differences between the results we observe and the results we expect to observe. Here are some examples of typical null hypotheses:

H_0: The seedlings in groups *A* and *B* do not differ in height (or the addition of nutrients does not alter seedling growth rates).

H_0: The seed color of offspring does not differ from the expected ratio of 3:1.

H_0: There is no relationship between hermit crabs' weight and the size of the shells they occupy in the field.

H_0: Juniors did not do better than sophomores on the midterm examination.

H_0: Activation of a particular intracellular enzyme does not alter neuronal survival rate in cell culture.

H_0: The amount of the steroid corticosterone released into the blood by starlings in response to stress does not vary seasonally.

It may seem surprising that the hypothesis to be tested is the one that anticipates no unusual effects; why bother doing the study if we begin by assuming that our treatment will be ineffective, or that there will be no differences in eye color, or that wing lengths will not differ from population to population? For one thing, the null hypothesis is chosen for testing because scientists must be cautious in drawing conclusions. **Hypotheses can never be proved; they can only be discredited or supported.** And the strongest statistical tests are those that discredit null hypotheses. Therefore, the cautious approach in testing the effect of a new drug is to assume that it will not cure the targeted ailment. The cautious approach in testing the effects of different diets on the growth rate or survival rate of a test organism is to assume that all diets will produce equivalent growth or survival—that is, that one diet is not superior to the others being tested. The cautious approach in testing the effects of a pollutant is to assume that the substance is not harmful. Only if we can discredit the null hypothesis (the hypothesis of no effect) can we tentatively embrace an alternative hypothesis—for example, that a particular drug *is* effective, or that wing lengths *do* differ among populations, or that a pollutant *is* harmful.

Please understand that **evidence against the null hypothesis does not *prove* that it is incorrect**. Neither does it *prove* that an alternative hypothesis is correct. Similarly, **failure to reject the null hypothesis does not *prove* that the null hypothesis is correct** or that an alternative hypothesis is incorrect. Therefore, **we must choose our words with great care when reporting the results of statistical tests**. This is a tricky business, which I will try to clarify in the next section.

Conducting the Analysis and Interpreting the Results

Once we have established our null hypothesis and collected the data for our study, statistical analysis of the data can begin. A large number of statistical tests have been developed, including the familiar chi-square test, the Student's t-test, and tests for correlation. The test that should be used to examine any particular set of data will depend on the type and amount of data collected and on the nature of the null hypothesis being addressed. If you are asked to conduct a statistical analysis of your data, your laboratory instructor will undoubtedly specify the test for you. Once the appropriate test is chosen, the data are maneuvered through one or more standard, prescribed formulas to calculate the desired test statistic. This test statistic may be a chi-square value, a t-value, an F-value, or any of a variety of other values associated with different tests; in all cases, the calculated test value will be a single number, such as 0.93 or 129.8. A calculated value close to 0 suggests that the data from the experiment are consistent with the null hypothesis (little deviation from the outcome that would be expected if the null hypothesis were true). A value very different from 0 indicates that the null hypothesis may be wrong, because the data obtained are very different from those that would be expected if the null hypothesis were true.

Returning to our seedling experiment, we wish to determine if adding certain nutrients alters seedling growth rates; the null hypothesis states that the nutrients have no effect. The appropriate test for this hypothesis is the t-test. Applying the formula provided in statistics textbooks, the value of the t-statistic calculated for the data obtained in our tomato seedling experiment turns out to be 1.62 (or -1.62; the sign makes no difference). This particular value has some probability of turning up even if the null hypothesis is true. Here, the argument gets a bit tricky. If we repeated the experiment exactly as before, using another set of 60 seeds, we would most likely obtain a somewhat different result and the t-statistic would have a different value, even though the null hypothesis might still be true. If we did 5 identical experiments, we would probably calculate 5 different t-values from the data. In other words, a statistic may take on a broad range of values even if the null hypothesis is correct, and each of these values has some probability of turning up in any single experiment. But some values are more likely to turn up than others.

Suppose the null hypothesis stating that adding nutrients does not alter the growth of tomato seedlings over the first 10 days of observation is actually correct. If we ran our experiment (with 30 seeds planted in each of the 2 treatment groups) 100 times, we might actually find no measurable difference between the average heights of the seedlings in some of the experiments; our calculated t-values for these data would then be 0. In most of the experiments, we would probably record small differences between the average sizes of seedlings in the 2 populations (and, for each of these experiments, calculate a t-value close to 0), and in a few experiments, purely by chance, we would probably record large differences (and calculate t-values either much larger or much smaller than 0). All these results are possible if we do enough experiments, even though the null hypothesis is correct, simply because the growth of seedlings varies even under a single set of experimental conditions. The oddball result may not come up very often, but there is always some probability that it will pop up in any given experiment. If we do only one experiment, we have no way of knowing whether we got an odd result or how odd our result is.

The important point here is that the outcome of an experiment or study can vary quite a lot, regardless of whether or not the null hypothesis is correct. A nonbiological example may help clarify this point: In coin tossing, a fair coin should, on average, produce an equal number of heads and tails. Yet experience tells us that 10 tosses in a row will often produce slightly more of one result; every now and then, we will actually end up tossing 10 heads in a row or 10 tails in a row, even though the coin is perfectly legitimate. Neither of these results will occur very often, but each will occur eventually if we repeat the experiment enough times.

Yes, the fact of the matter is that organisms in the real world show considerable morphological, physiological, and behavioral variability, and that the only way to know, with certainty, that our one experiment is a true reflection of that world is to measure or count every individual in the population under consideration (e.g., plant every tomato seed in the world, and then measure every seedling after 10 days) or conduct an infinite number of experiments. This is not a practical solution to the problem. The next best alternative is to use statistical analysis. Statistics cannot tell us whether we have revealed THE TRUTH, but they can indicate just how convincing our results are (or aren't), and they can guide the direction of future studies.

The numerical value of any calculated test statistic has some probability of turning up when the null hypothesis is true. Statisticians tell us, for example, that values of t are distributed as shown in Figure 7, and that values of chi-square (χ^2) are distributed as shown in Figure 8. **If the null hypothesis is correct, values of each statistic will usually fall within a**

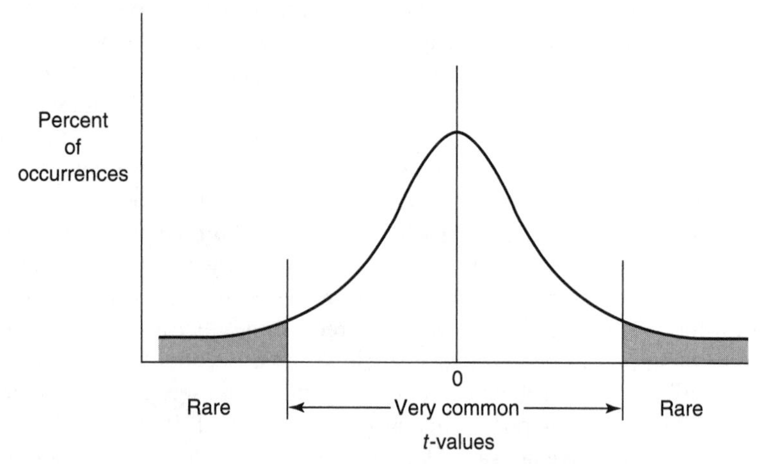

Figure 7. The distribution of t-values expected when the null hypothesis (H_0) is true. The exact shape of the distribution depends on the number of degrees of freedom (p. 60), which increases with sample size. A wide range of values may occur, but some values will occur more commonly than others. Obtaining a common value for t causes us to provisionally accept H_0. Obtaining a rare value for t causes us to doubt the validity of H_0.

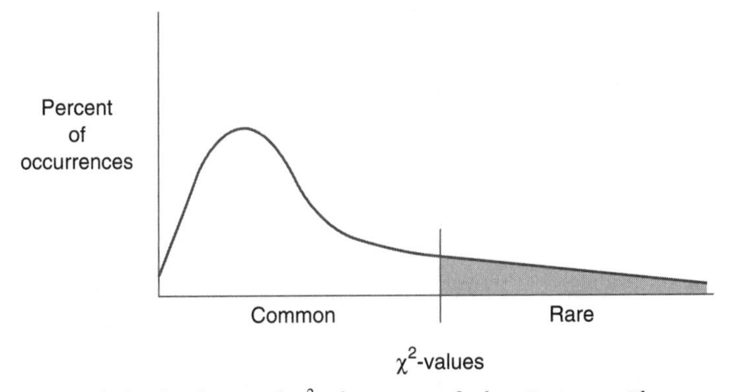

Figure 8. The distribution of χ^2-values expected when H_0 is true. The exact shape of the distribution depends on the number of degrees of freedom. A wide range of values may occur, but some values will occur more often than others. The rarer the value obtained, the less confidence we can have in the validity of H_0.

certain range, as indicated; these values will have the greatest probability of turning up in any individual experiment. If the *t*-value calculated for our experiment falls within the range indicated as "very common," we are probably safe in accepting the null hypothesis; at least we have no reason to disbelieve it. Even if the null hypothesis is correct, however, unusual values of *t* or of χ^2 will occasionally occur. We are, after all, randomly picking only a few seeds to plant out of a bag that may contain many thousands of seeds; it could be just our bad luck to have picked mostly those seeds that are most unlike the average seed.

If we calculate an unusual (very far from 0) value for *t* using the data from our experiment, how can we decide to reject the null hypothesis when we know there is still some small chance that H_0 is correct and that we have simply witnessed a rare event? Well, we must admit that we are not omniscient. **We must be willing to take a certain amount of risk in drawing conclusions from our data; the amount of risk being taken can be specified.** Typically, researchers assume that if their unusual—that is, far from 0—value of *t* (or of some other statistic) would turn up fewer than 5 times in 100 repetitions of the same experiment when the null hypothesis were true, then this oddball value of *t* is a strong argument *against* the null hypothesis being correct; H_0 is then tentatively rejected. That is, the large calculated value of *t* would be so rarely encountered if the null hypothesis were true that the null hypothesis is *probably* wrong. However, there is still the 5 in 100 chance that the null hypothesis is correct and that the researchers, through random chance, happened upon atypical results in their experiment. Tossing 10 heads in a row using a fair coin won't happen very often, but it *will* happen eventually. Tossing 100 heads in a row is an even rarer event, but it could happen. If you conducted only one tossing experiment of 100 flips and tossed only heads, you could tentatively reject the null hypothesis that the coin is fair, but you wouldn't know with certainty that you were correct. Would you bet your car, your savings account, your little finger, or your stereo that you would get all heads if you did another round of 100 tosses? Only if the coin has 2 heads.

Sometimes researchers will be even more cautious in rejecting H_0 and reject it only when there is less than a 1 in 100 chance of doing so incorrectly. Then, you will need a very unusual value of t indeed (unusual if H_0 is true) before rejecting the null hypothesis. The downside, of course, is that by making it harder to reject H_0, you are also making it easier to accept H_0 when, in fact, H_0 is wrong. There is no win–win situation available in this game.

It is, I hope, becoming clear why experiments must be repeated many times, and with much replication within an experiment, before the results become convincing. Such is the challenge of doing biology. **Only when large values of a test statistic appear many times can we become fully confident that the null hypothesis deserves to be rejected.** Only when low values appear many times can we become confident that the null hypothesis is most likely correct.

Degrees of Freedom

The exact shape of statistical distributions like those shown in Figures 7 and 8 varies slightly with something called "degrees of freedom" (abbreviated "d.f."), which is related to sample size and/or the number of treatments included in the study. As a simple example, consider a t-test comparing the means of 2 columns of numbers (e.g., the final heights of seedlings with and without added nutrients, or the number of moles of product formed after 5 minutes of an enzymatic reaction at 2 temperatures). Suppose we have 10 replicates in one column and 9 replicates in the other—somebody dropped one test tube and lost that sample. For the first column of data, if we know 9 of the numbers and the mean of all 10, then the 10th number has no freedom to vary—knowing the mean locks that last number into a particular value. Thus, each time we calculate a mean, we lose one degree of freedom. In calculating 2 means, we lose 2 degrees of freedom; in our example, then, the number of degrees of freedom is $(10 + 9) - 2 = 17$. For the data shown in Figure 5, the number of degrees of freedom $= (26 + 23) - 2 = 47$. The important point is that the exact shapes of the t and χ^2 distributions shown in

Figures 7 and 8 (and those for other statistics as well) are different for different degrees of freedom. Thus, the cutoff between common and unusual values for any statistic if H_0 is true differs among experiments with different sample sizes. As discussed earlier, in using statistical tables to determine the numerical cutoff value, you need to know the number of degrees of freedom in your analysis as well as the degree of risk you are willing to assume for incorrectly rejecting H_0.

SUMMARY: USING STATISTICS TO TEST HYPOTHESES

In performing a statistical test, you first decide on a reasonable degree of risk (usually 5% or less) of incorrectly rejecting the null hypothesis (and incorrectly accepting an alternative hypothesis). Then, you perform the study, plug the data into the appropriate formula to calculate the value of the appropriate statistic, and end up with a single number. You then look in the appropriate statistical table (using the correct number of degrees of freedom) to see whether this number is within the range of values expected when the null hypothesis is correct. If your number lies within the expected range of values, your data support (but do not prove) the null hypothesis. If your number lies outside the range of commonly expected values, your data do not support H_0; they support (but do not prove) the alternative. Remember, there is always some small chance that you are making the wrong decision by rejecting H_0. Similarly, if your number lies within the range of commonly expected values, there is always some chance that you are making the wrong decision by accepting H_0. For this reason, our data cannot *prove* any particular hypothesis; they can only either favor or argue against the null hypothesis.

MOVING BEYOND *p*-VALUES

I have focused on hypothesis testing because it dominates the research literature in so many fields. The past 15 or 20 years, however, have seen some disillusionment with the value of hypothesis testing and the reporting of probability values (p-values). Small p-values are strong evidence against H_0, but as discussed earlier, it's hard to know what to make of large p-values, especially those greater than about 0.1. Such values certainly don't give you any reason to reject H_0, but neither do they necessarily mean that you should accept it—basically, high p-values mean only that you must withhold judgment.

Even small *p*-values can be difficult to interpret. For example, if sample sizes are large enough, perhaps hundreds or thousands of individuals in each treatment, then even very, very small differences between means will turn out to be statistically significant.

Thus, many researchers are now providing additional information when reporting their results. In particular, it is now becoming common to also report "statistical power" and effect magnitudes, as described below.

Statistical Power

Often, the practical design of an experiment makes it hard to reject H_0 even when it's false. Because of the large variation within samples, for example, you might simply have not included enough replicates in your study, or enough individuals per replicate, to document the real differences that exist between the means of 2 or more populations, or to uncover a correlation between 2 variables. It is slowly becoming more common for authors to report the "statistical power" of the tests they conduct—particularly when the study fails to reject H_0—along with the *p*-values they obtain, as in this example:

> Blood prolactin concentrations in male blackbirds provisioning their young were not significantly related to brood size ($r^2 = 0.03$, $n = 11$, $p = 0.595$, power = 0.15).*

In other words, there was no significant relationship between brood size and the hormonal levels of the males visiting those broods ($p = 0.595$), but the power of the test to detect such differences was very low. Either there really is no difference, or there is a difference that we are unable to detect because of small sample size. We just don't know. By the way, r^2 is a measure of correlation; in particular, it represents the amount of variation (3% in this case) in *Y* that is explained by variation in (see also p. 66).

In contrast, a test resulting in the same high *p*-value but with high statistical power suggests that we can have greater confidence in accepting H_0. In the Discussion section of your paper, then, you would be able to write something like this, if you had conducted a power analysis:

> The large number of replicates used in this study (33–39) gave the study sufficient power (0.79) to detect an effect of leaf damage on leaf palatability to insects. The fact that we found no such effect suggests, then, that leaf damage did not induce chemical defenses against this insect, at least not over an 8-day period.

---

*Based on Préault et al., 2005. *Behav. Ecol. Sociobiol.* 58: 497–505.

Thus, "power" is basically the probability of correctly detecting a false H_0—that is, it is the probability of ***not* accepting H_0 when it is wrong**—and "1 – power" is the probability of accepting H_0 when it is wrong. In the example given above with the male blackbirds, there is an 85% chance [(1.0 – 0.15) ∞ 100%)] of not rejecting H_0 when it is, in fact, wrong.

By convention, researchers are happy if they make the right decision 80% of the time; that is, they are usually happy if their statistical tests have a power of 0.8.

Power analysis is commonly used in designing studies. For example, you could ask, "How many replicates of 12 animals each will I need in my study to detect a 10% difference between means?" Although these estimates are difficult to calculate by hand, software for making the calculations is becoming widely available.

Effect Magnitudes and Alternative Analyses

When is a difference a meaningful difference, biologically or medically? Suppose you determine a significant difference ($p < 0.05$) between the means of a control and an experimental population, but the difference between means is only 1.2%. Now, suppose instead that you determine a significant difference between the means of a control and another experimental population ($p < 0.05$), but that difference between means is 26%. Clearly, these two results are not equivalent even if the *p*-values are identical.

Many researchers now report effect magnitudes when they present their data in the Results section. For example, "The increase in volumetric bone mineral density at the trabecular spine was greatest in the parathyroid hormone–alendronate group (31%, $p < 0.001$)"[*] and "Rock lobsters were 3.8 times more abundant per m^2 at the protected sites than at the non-protected sites ($\chi^2 = 54.5, p < 0.001$)."[**]

Individual studies often lack significant power to detect a meaningful difference, even when one exists, because of necessarily small sample sizes, so many researchers are now doing **meta-analyses**, in which results of many individual studies, usually done by many different researchers, are combined into a single study. The results are still examined by hypothesis testing.

---

*From Black et al., 2005. *New Engl. J. Med.* 353: 555–565.

**Based on Langlois et al., 2005. *Ecology* 86: 1508–1519.

Sometimes you will run across a very different approach to evaluating data, one that does not involve hypothesis testing. Instead, the experimenters compare their data with those predicted by selected mathematical models and then evaluate how well the data fit, or do not fit, the models. This general approach is referred to as **Bayesian**, and a typical way of evaluating the data is called **Maximum Likelihood**. Again, you don't need to understand the details of how the calculations were made in order to understand the point of the study: The data either do or do not conform to the predictions of the model.

READING ABOUT STATISTICS

When you read the results of statistical analyses in a published research paper, you don't need to understand how the particular calculations were made in order to understand what the results mean. **A small *p*-value means that if the null hypothesis were true, we would find a difference as large as the one found in that study very rarely.** For example, "$p < 0.001$" means that such a large difference would be expected by chance fewer than 1 time in 1,000 repeats of the study if H_0 were true. Thus, the null hypothesis is probably wrong. **A large *p*-value means that the researchers obtained a difference of about the size expected if the null hypothesis were true.** For example, "$p = 0.28$" means that the results would turn up at least 28 times out of 100 repeats of the experiment if H_0 were true. Thus, the researchers can't reject H_0 with confidence, especially if the statistical power of the test is low; instead, they must retain it for the time being as part of a work in progress.

WRITING ABOUT STATISTICS

If you have analyzed your data using appropriate statistical procedures, the products of your heavy labor are readily and unceremoniously incorporated into the Results section of your report to support any major trends that you see in your data, as in the following 3 examples. **Note that in Example 1 below, the author reports the sample size ($N = 30$ caterpillars), the test statistic and its value, the number of degrees of freedom associated with the test, and the size of the associated *p*-value.** Note also that the writer focuses on the biology being studied—that is, on the biological question being addressed—rather than on the statistics themselves.

EXAMPLE 1

For 30 caterpillars reared on the mustard-flavored diet and subsequently given a choice of foods, the caterpillars showed a statistically significant preference for the mustard diet ($\chi^2 = 17.3$; *d.f.* = 1; $p < 0.05$). For 30 caterpillars reared on the quinine-flavored diet, however, there was no significant influence of previous experience on the choice of food ($\chi^2 = 0.12$; *d.f* = 1; $p > 0.10$).

In this example, H_0 states that prior experience does not influence the subsequent choice of food by caterpillars; "$p < 0.05$" means that if we were to conduct the same experiment 100 times and H_0 were true, such a high value for χ^2 would be expected to occur in fewer than 5 of those 100 studies. In other words, the probability of making the mistake of rejecting H_0 when it is, in fact, true is less than 5%. You can therefore feel reasonably safe in rejecting H_0 in favor of the alternative: that prior experience did influence subsequent food selection by caterpillars reared on the mustard diet.

Different results were obtained, however, for the caterpillars reared on the quinine-flavored diet; "$p > 0.10$" means that if the experiment were repeated 100 times and H_0 were true, you would expect to calculate such a small value of χ^2 in at least 10 of the 100 trials. In other words, the probability of getting this χ^2-value with H_0 being true is rather high; certainly, the χ^2-value is not unusual enough for you to mistrust H_0 and run the risk of rejecting the null hypothesis when it might, in fact, be true. **This does not mean that H_0 is true, only that we do not have enough evidence to reject it.** The quinine-flavored diet may have altered dietary preference—the variability in response may simply have been too great for us to perceive the effect with the small number of caterpillars used in our study. It is also possible, of course, that the diet on which caterpillars were fed affected food choices in ways that we did not test for. Thus, the wording in Example 1 was very carefully chosen to say no more than is safe to say.

EXAMPLE 2

Over the first 10 days of observation, growth of seedlings receiving the nutrient supplement was not significantly faster than the growth of the seedlings receiving only water (2-sample *t*-test, $t = 1.62$; *d.f.* = 47; $p = 0.11$).

In this second example, the writer has chosen to report the actual p-value rather than writing "$p > 0.10$." The null hypothesis (H_0) states that the nutrient supplement does not influence plant growth; "$p = 0.11$" means that if the experiment were repeated 100 times and H_0 were true, you would expect to calculate such a low value of t in 11 of the 100 trials. As before, you have obtained a value of t that would be fairly common if H_0 were true, so you have no reason to reject H_0. It is, of course, possible that H_0 is actually false and the nutrients really do promote seedling growth, and that you just happened upon an unusual set of samples that gave a misleadingly small t-value. If such is the case, repeating the experiment should produce different results and larger t-values. But with only the data before you, you cannot reject H_0 with confidence.

EXAMPLE 3

The hermit crabs in our sample ($N = 12$) showed a significant relationship between their wet weights and the size of the shells they occupied in the field ($r^2 = 0.477$; test for zero slope: $F = 9.104$; $d.f. = 1, 12$; $p = 0.013$).

In this case, hermit crab weight and shell size were correlated (the slope of the line relating the 2 variables differs significantly from zero), even though variation in hermit crab weight accounted for only 47.7% of the variation in the sizes of the shells occupied (as shown by the r^2 term).

Note carefully how the word *significant* was used in the above examples, **and that the writers put the focus clearly on biology**; they said very little about the statistics themselves. Statistics are used only to support any claims you wish to make about your results, as in examples 1–3; resist the temptation to ramble on about how the statistics were calculated, how brilliant you are to have figured out which calculations to make, or how awful it was to make those calculations.

Compare the first 3 examples with the following 2 examples.

EXAMPLE 4

Wrong:

A chi-square value of 6.25, with a p of 0.0124, revealed that the physical condition of the shell had a significant influence on shell choice by the 15 hermit crabs used in our study.

Corrected:

The physical condition of the shell had a significant influence on shell choice by the hermit crabs ($\chi^2 = 6.25$; $d.f. = 1$; $N = 15$ hermit crabs, $p = 0.0124$).

EXAMPLE 5

Wrong:

A Student's 2-sample *t*-test was used to determine the significance of the difference in mean interaction times. The data were not significant (i.e., we found no significant results in our experiment), and there was no difference in the mean contact time between hermit crabs in the presence or absence of predators.

Analysis of Example 5:

The first sentence, of course, should not be there at all. With the second sentence, the student has fallen into the very common trap of confusing the results of a significance test with the value of the data, or of the study itself ("The data were not significant . . . "). There is nothing wrong with the data. **Don't apologize for results that support H_0. Failure to discredit H_0 does not mean that your experiment was a failure.** If your sample size was small and the amount of natural variability was high, then it is very hard to discredit a null hypothesis even when H_0 is wrong. And, of course, H_0 might actually be correct.

The student in this example then uses the results of a significance test to make a definitive pronouncement ("there was no difference in . . . "), by omitting the word *significant* from the sentence. As discussed earlier, statistics do not prove things; they only indicate degrees of likelihood. "No significant difference" is not the same as "No difference." Finally, the student provides no statistical support for the final statement made.

Corrected:

The presence of a predator did not significantly affect the amount of time that hermit crabs spent interacting with each other (2-sample *t*-test, $t = 1.012$; $d.f. = 12$; $p = 0.332$).

For more detailed advice about discussing "negative" results (i.e., those that fail to reject H_0) see Chapter 9, on writing a Discussion section (p. 192).

As mentioned earlier (pp. 62–64), it is becoming common to report not just *p*-values but also statistical power and effect sizes to help interpret the reported *p*-values. Examples are given on pages 62–63. Ask your instructor whether he or she expects you to include this information in your reports.

SUMMARY

1. Individuals typically differ in size, physiology, and behavior, making it difficult to definitively characterize the average (mean) trait without large sample sizes.
2. Variation in traits is typically characterized by variance, standard deviation (SD), or standard error of the mean (SEM)—the larger the numbers, the greater the amount of variation among measurements—and by confidence intervals.
3. Null hypotheses (H_0) assume that treatments have no effect, that there are no differences between groups, that there is no correlation between the traits of interest; the null hypothesis is rejected only when statistical testing makes it seem very unlikely.
4. Rejecting the null hypothesis does not prove that it is wrong; accepting the null hypothesis does not prove that it is correct.
5. When incorporating the results of statistical tests into reports, emphasize the biological result, not the statistics; that is, use statistics in the same way that you use references to support statements of fact or opinion:

   > There was no effect of the nutrient supplement on rate of tomato seedling growth (2-sample *t*-test, $t = 1.62$, $d.f. = 47$, $p > 0.10$).
   >
   > The nutrient supplement caused significantly more rapid plant growth in tomato seedlings (2-sample *t*-test, $t = 3.81$, $d.f. = 47$, $p < 0.01$).

5

CITING SOURCES AND LISTING REFERENCES

What Lies Ahead? In This Chapter, You Will Learn

- To cite references by author name and year of publication
- How to determine what needs to be cited and how to cite your references concisely
- How to prepare a valid Literature Cited section
- The importance of using a consistent style in listing references

As described briefly in Chapter 1, **all statements of fact and opinion require support** to be convincing to the thoughtful, critical reader. The firmer the statement and the more important it is to your argument, the greater the need for support. In research reports (including lab reports), review papers ("term papers"), and theses, factual statements are often supported by reference to the source (or sources) of the facts presented. Therefore, in a separate section at the end of your manuscript, you must list the books, research articles, and Web sites referred to so that they can be located by interested (or skeptical) readers.

In general, your instructor will expect you to cite textbooks, research articles, and review papers published in "peer-reviewed" journals (see p. 27). Articles from some magazines and newspapers may also be acceptable in introductory courses, but you should check with your instructor before including such citations.

CITING SOURCES

Here are a few general rules to follow when citing sources to back up factual statements. These rules apply to review papers, essays, theses, and research reports and to citations throughout the research report:

1. **Don't footnote: Cite by author and year of publication.** In most published research papers, references are cited

directly in the text by author and year of publication, as in the following example:

> A variety of organic molecules are commonly used to maintain or adjust the osmotic concentration of intracellular fluids (Hochachka and Somero, 1984; Schmidt-Nielsen, 1990).

Note that the period, both here and in the examples that follow, appears at the end of the closing parenthesis since the reference, including the publication date, is part of the sentence.

- Where appropriate, you may instead **incorporate the authors' names directly into a sentence**:

> Kim (1976) demonstrated that magnetic fields established by direct current can alter the rates of enzyme-mediated reactions in cell-free systems.

or

> The ability of magnetic fields established by direct current to alter the activity of certain liver enzymes was first demonstrated by Kim (1976).

- Try to **make the relevance of the cited reference clear** to the reader. For example, rather than writing:

> Temperature tolerances have been determined for gastropods, bivalves, annelids, and insects (Merz, 1988; Heibert Burch, 1998; Merz and Heibert Burch, 1993 a, b).

it would be clearer to write:

> Temperature tolerances have been determined for gastropods (Merz, 1988), bivalves and annelids (Heibert Burch, 1998), and insects (Merz and Heibert Burch, 1993 a, b).

- When **more than 2 authors** have collaborated on a single publication, a shortcut is standard practice:

> A mutation is defined as any change occurring in the nitrogenous base sequence of DNA (Tortora *et al.,* 1982).

The *et al.* is an abbreviation for *et alii*, meaning "and others." The words are underlined or italicized, even when abbreviated, because they are in a foreign language, Latin; underlining tells a printer to set the designated words or letters in italics. Note that there is no period after the "*et*"; it means "and."

- If you cite **2 papers published in a single year by the same author**, use letters to distinguish between them: (Dopman and Li, 2008 a, b) or (Ellmore *et al.*, 2009 a, b).
- You can **cite your laboratory manual** by its author (e.g., Professor S. Heibert Burch, 2010) or as follows:

  Preparation of buffers and other solutions is described elsewhere (Biology 1 Laboratory Manual, Swarthmore College, 2010).

- If your information comes from a **lecture or from a conversation** with a particular individual, support your statement as follows:

  California gray whales migrate up to 18,000 km yearly (Professor Tim Rawlings, personal communication, September 2010).

- **In some biological journals, the "author-year" format for citing references has been replaced by a more compact "number-sequence" format.** In the number-sequence format, each reference cited in the paper is represented by a unique number. The first paper to be cited is assigned the number 1, the second paper to be cited is assigned the number 2, and so forth, as in the following example:

  Substantial declines in amphibian populations have been documented from numerous locations around the world over the past 40 years (3,5,8–12).

The numerals 3 and 5 represent the 3rd and 5th references cited in the student's paper. Earlier in the paper, the author must have cited references 1 through 7. References 3 and 5, although cited earlier in the paper, are relevant here and so are cited again using the same numbers.

Unless your instructor asks you to adopt the number-sequence format in your own work, use the author-year format discussed earlier. For one thing, it is a nuisance for readers to have to keep turning to the back of a paper to see whose work is being cited. Moreover, biologists commonly refer to particular papers by their author(s) and year of publication ("Have you read Booth and Gable, 2008, yet?"), and using the author-year format of citing references helps you remember who did what, and when. But perhaps most important, research papers are written by real people, and as you become more familiar with the literature in any particular field, you will find yourself coming across many of the same names repeatedly. Using the author-year format is a good way to learn which people are best associated with your field of interest. Perhaps you will decide to do graduate work with one of these people in the future.

2. **Be concise in citing references.** Avoid writing:

   In his classic work *The Biology of Marine Animals,* published in 1967, Colin Nicol reviewed the literature on invertebrate bioluminescence.

   Instead, write:

   The phenomenon of invertebrate bioluminescence was carefully reviewed by Nicol (1967).

   Again, the period follows the parenthesis.

3. **Cite only those sources you have actually read and would feel confident discussing with your instructor.** Don't list references simply to add bulk to this section of your report; your instructor is perfectly justified in expecting you to be able to discuss any material you cite. Listing a few references you have thoughtfully incorporated into your paper should do more for your grade than an attempt to create the illusion that you have read everything in the library.

   You may occasionally have to cite a source that you have not actually read. For example, results reported by Hendler (1999) may be cited in a book or article written by Dufus (2003), and you have read only the work by Dufus. Your citation should then read "(Hendler, 1999, as cited by Dufus, 2003)." Let Dufus take the blame if he or she has misinterpreted something. In the Literature Cited section of your report, you would include both sources.

4. If **citing a review paper that does not include the original data supporting a particular statement** that you wish to make, cite the reference as " … (reviewed by Furhman, 2004)."

5. **Avoid citation overkill.** When discussing a series of facts from a single source or group of sources, it is not necessary to cite the same source(s) in every sentence. There are many ways of informing the reader that a series of sentences is based on a single source of information, as in the following example:

   Hochachka and Somero (1984) discuss the physiological adaptations of diving mammals in considerable detail.

> In particular, they note that diving Weddell seals exhibit a pronounced decline in both rate of metabolism and ATP turnover rate. In one experiment, ATP turnover rates were reduced by as much as 50% during a 20-minute dive.

6. When citing an online journal, use the same format (author, year of publication) as for a hard-copy journal.
7. Don't cite personal Web sites or any other Web sites (e.g., Wikipedia) that publish information without verifiable authoritative review.

   As explained in Chapter 2 (pp. 27, 32), information posted on Web sites is ephemeral and has usually not been peer-reviewed: Avoid using Web pages as sources of information unless you are fully confident of the accuracy of the material presented. In general, this means relying only on peer-reviewed electronic journals or Web sites maintained by recognized scientific authorities, such as those associated with major museums and research institutions or by government organizations, such as the U.S. Department of Agriculture (USDA), the National Oceanic and Atmospheric Administration (NOAA), and the World Health Organization (WHO). Such Web sites will include ".edu", ".gov", or ".org" in their URLs. If you have any questions about whether a Web site is valid, ask your instructor. Wikipedia articles are often good sources of valid reference material; you should read and cite those sources, but not the Wikipedia article itself. **Never insert wording taken directly from a Wikipedia article into your papers**; that would be plagiarism (pp. 13).

SUMMARY OF CITATION FORMAT RULES

- **Use the author-year format** for citing references unless you are told otherwise by your instructor or unless you are submitting a manuscript to a journal using a different (number-citation) format.
- **Cite authors only by their last names** unless you include in your paper citations by 2 authors sharing a last name (e.g., Bilbo Baggins and Frodo Baggins). In such a case, distinguish between the 2 authors by using the first letter of the first name (e.g., B. Baggins, 1946).
- Cite **work by 2 authors** using the last names of both (e.g., Fraga and Iyengar, 2000).

- Cite **work by 3 or more authors** using only the last name of the first author, followed by *et al.* (meaning, "and others").
- **If you must cite a reference that you have not read**, do it as follows: (Tankersly, 1995, as cited by Rittschof, 2006).
- Cite information provided directly by your instructor (orally or through email) as follows: (J. Bolker, personal communication, Oct. 2011).

PREPARING THE LITERATURE CITED SECTION

Whenever you cite sources to support statements, you must provide a separate Literature Cited section, giving the full citations for each source cited. This presentation enables the interested reader—including, perhaps, you, at a later date—to locate and examine the basis for factual statements made in your report. It occasionally happens that a reference is used incorrectly; your interpretation or recollection of what was said in a textbook, lecture, or journal article may be wrong. By giving the source of your information, the reader can more easily recognize such errors. If the reader is your instructor, this list of references may provide an opportunity for him or her to correct any misconceptions you have acquired. If you fail to provide the source of your information, your instructor will have more difficulty in determining where you went wrong. Proper referencing is even more crucial in scientific publications. Misstatements of fact are readily propagated in the literature by others; the Literature Cited section of a report enables readers to verify all factual statements made. The careful scientist consults the listed references before accepting statements made by other authors.

Listing the References—General Rules

Include only those references that you have actually read (see p. 72 for one exception to this rule) **and that you specifically mention in your report or paper, and include all of the references that you cite**. Unless you are told otherwise by your instructor, list references in alphabetical order according to the last name of the first author of each publication. If you cite several papers written by the same author, list those papers chronologically. If one author has published 2 papers in the same year, list those papers as, for example, "Hentschel, B. 1995a," and "Hentschel, B. 1995b."

Each listing must include the names of all authors, the year of publication, and the full title of the paper, article, or book.

In addition, **when citing books**, you must report the publisher, the place of publication, and the pages referred to or the total number of pages in the book.

When citing journal articles, you must include the name of the journal, the volume number of the journal, and the page numbers of the article consulted.

When listing an article from a hard-copy journal that you accessed online, use the standard hard-copy journal format; there is no need to include a URL, for example.

When citing exclusively online journals and other Internet sources, you must include the date that the material was posted (or the most recent revision date), the date that you accessed the material, the full URL for the Web site, and an indication (in brackets: []) that the medium is the Internet. See pp. 77–79 for examples of correct citations. Standards are still being established; with Internet sources, it is sometimes difficult, or impossible, to determine the author of the material, the publication date, the publisher, and even the title of the article! The most recent and detailed advice for citing Internet sources is available through the Council of Science Editors and the National Library of Medicine (see Appendix C for the Web sites).

Listing the References—Using the Correct Format

Here are a few general rules followed by most journals:

- List references in alphabetical order and then chronologically.
- Spell out only the last names of authors; initials are used for first and middle names.
- Include the names of all authors, even though the names of only 1 or at most 2 authors (e.g., Croll and Voronezhskaya, 1996; Woodin *et al.*, 1995) are cited in the text of the report.
- Latin names, including species names, are italicized, or they are underlined to indicate italics.
- Titles of journal articles are not enclosed within quotation marks, and only the first word is capitalized.
- For book titles, capitalize the first letter of each word other than *the*, *and*, or *or*.
- For books, indicate the total number of pages in the book if you are citing the entire book, or indicate the specific page numbers if you are citing a particular chapter in the book or a particular section of the book.
- Journal names are usually abbreviated. In particular, the word *Journal* is abbreviated as *J.*, and words ending in *-ology* are

usually abbreviated as *-ol.* The *Journal of Zoology* thus becomes *J. Zool.* Do not abbreviate the names of journals whose titles are single words (e.g., *Science* or *Evolution*). Acceptable abbreviations for the titles of journals can usually be found within the journals themselves.

Unfortunately, formats differ from journal to journal, despite the best efforts of the Council of Science Editors. Consider, for example, how one particular citation would be formatted for 4 different journals:

For Biological Bulletin	**Contakos, S.P., C.M. Gaydos, E.C. Pfeil, and K.A. McLaughlin. 2006.** Subdividing the embryo: A role for Notch signaling during germ layer patterning in *Xenopus laevis. Devel. Biol.* **288**: 122–134.
For Developmental Biology	Contakos, S.P., Gaydos, C.M., Pfeil, E.C., McLaughlin, K.A., 2006. Subdividing the embryo: A role for Notch signaling during germ layer patterning in *Xenopus laevis.* Dev. Biol. 288, 122–134.
For Ecology	Contakos, S.P., C.M. Gaydos, E.C. Pfeil, and K.A. McLaughlin. 2006. Subdividing the embryo: A role for Notch signaling during germ layer patterning in *Xenopus laevis.* Developmental Biology **288**:122–134.
For Science	Contakos, S.P., C.M. Gaydos, E.C. Pfeil, K.A. McLaughlin. *Devel. Biol.* **288**, 122 (2006).

List all the differences you can find! Why do citation formats differ so much among journals? I have no idea, and I wish they didn't. But they do. The following examples should help you prepare your own Literature Cited sections, but check to see if your instructor wants you to follow a particular format, or the format used by a particular journal.

The most important rule in preparing the Literature Cited section is to provide all the information required and to be consistent in the manner in which you present it. If you are using bibliographic management software (see Technology Tip 3 below), this can be specified at the touch of a button.

When preparing a paper for publication, follow the format used by the journal to which you are submitting the paper, and follow it to the last detail.

Note in the following examples that the citation begins at the far left, and that subsequent lines are indented several spaces to the right. This format is called a "hanging indent" (see Technology Tip 4, p. 77, to learn how to do this automatically in Microsoft Word). A sample Literature Cited section appears at the end of this chapter.

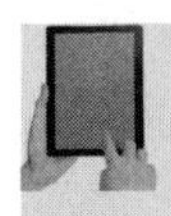

TECHNOLOGY TIP 3

Bibliographic management software

As you start accumulating large numbers of references, you might want to start putting them into a bibliographic database manager, such as EndNote, Refworks, or Zotero. Zotero is free, and Refworks may be available for free through your college or university library. Bibliographic management software allows you to retrieve your references by keyword searching and to print your references in any desired format. Both EndNote and Refworks allow you to share reference listings with fellow students and other collaborators.

TECHNOLOGY TIP 4

Producing hanging indents

To produce the "hanging indent" format shown in the examples below:

- Open the Format menu in the Word tool bar.
- Click on Paragraph.
- Click on the term Special.
- Select "hanging" and specify how much you would like to indent.

Alternatively, you can type all the references first and then highlight them. Then choose the hanging indent option in the Format/Paragraph menus and specify the amount of indenting that you want.

Format for Journal References

Jarrett, J.N. 1997. Temporal variation in substratum specificity of *Semibalanus balanoides* (Linnaeus) cyprids. *J. Exp. Mar. Biol. Ecol.* 211: 103–114.

LeBlanc-Straceski, J., Sokac, M.A., Bement, W., Sobrado, P., Lemoine, L. 2009. Developmental expression of *Xenopus* Myosin 1d (XlMyo1d) and identification of a Myo1d tail homology that overlaps TH1. *Develop. Growth Differ.* 51: 443–451.

Woodin, S.A., Lindsay, S.M., Wethey, D.S. 1995. Process-specific recruitment cues in marine sedimentary systems. *Biol. Bull.* 189: 49–58.

Format for Book References

Nybakken, J.W., Bertness, M.D. 2005. *Marine Biology: An Ecological Approach*, 6th ed. Pearson Education, Inc., CA, pp. 25–31.

Format for an Article from a Book

Thompson, S.N. 1997. Physiology and biochemistry of snail-larval trematode relationships. In: *Advances in Trematode Biology* (Fried, B., Graczyk, T.K., eds.). CRC Press, New York, NY, pp. 149–195.

Format for a Laboratory Manual or Handout

Bernheim, H. 2002. Principles of physiology, using insects as models. II. Excretion of organic compounds by Malpighian tubules. Biology 50 Laboratory Handout. Tufts University, Medford, MA.

Biology 13 Laboratory Manual. 2005. Exercise in enzyme kinetics, pp. 16–23. Tufts University, Medford, MA.

Format for an article in a newspaper

Chang, K. 2011. Images of fossil birds show ancient pigments. New York Times, Section F:10 (col. 1).

Format for an article from a magazine

Losos, J.B. Evolution: A lizard's tale. Sci. Am. 2001 Mar; 284(3): 64–69.

Format for Items from the World Wide Web

Lawrence, R.A. A review of the medical benefits and contraindications to breastfeeding in the United States [Internet]. Arlington (VA): National Center for Education in Maternal and Child Health; 1997 Oct [cited 2005 Oct 27]. 40 p. Available from http://www.ncemch.org/pubs/PDFs/breastfeedingTIB.pdf

Note that there is no period at the end of the URL.

For the latest information on citing Web sources, see the Web sites listed in Appendix C.

A SAMPLE LITERATURE CITED SECTION

A sample Literature Cited section follows, with items arranged alphabetically and chronologically. Your instructor may specify a different format for this section of your report, so check first if you are uncertain.

Literature Cited

Journal, > 2 authors

Bayne, B.L., Livingstone, D.R., Moore, M.N., Widdows, J. 1976. A cytochemical and biochemical index of stress in *Mytilus edulis* L. *Mar. Poll. Bull.* 7: 221–224.

Journal, 2 authors

Finch, C.E., Rose, M.R. 1995. Hormones and the physiological architecture of life history evolution. *Q. Rev. Biol.* 70: 1–52.

Lab manual

Biology 220 Laboratory Manual. 2005. Exercise in enzyme kinetics, pp. 16–23. College of Wooster, OH.

Web journal

Fox, D.S., Heitman, J. 2002. Good fungi gone bad: the corruption of calcineurin. BioEssays [Internet] 2002 [cited 2005 Dec 18]; 24 (10): 894–903. Available from http://www3.interscience.wiley.com/cgi-bin/abstract/98518344/ABSTRACT

Book, 2 authors

Haas, W., Haberl, B. 1997. Host recognition by trematode miracidia and cercariae. In: *Advances in Trematode Biology* (B. Fried and T. L. Graczyk, eds.), CRC Press, New York, NY, pp. 197–227.

2 papers by same author in same year

McVey, M., LaRocque, J.R., Adams, M.D., Sekelsky, J.J. 2004a. Formation of deletions during double-strand break repair in *Drosophila* DmBlm mutants occurs after strand invasion. *Proc. Nat. Acad. Sci. USA* 101: 15694–15699.

McVey, M., Radut, D., Sekelsky, J.J. 2004b. End-joining repair of double-strand breaks in *Drosophila melanogaster* is largely DNA ligase IV independent. *Genetics* 168: 2067–2076.

Journal, 1 author

Orians, C. 2005. Herbivores, vascular pathways, and systemic induction: facts and artifacts. *J. Chem. Ecol.* 31: 2231–2241.

Book, 4 authors

Purves, W.K., Sadava, D., Orians, G.H., Heller, H.C. 2000. *Life: The Science of Biology,* 6th ed. Sinauer Assoc., Sunderland, MA, pp. 374–379.

Book, 2 authors

Quinn, G.P., Keough, M.J. 2002. *Experimental design and data analysis for biologists.* Cambridge University Press, New York, NY, 537 pp.

Web site

Wray, G.A. Echinodermata. [Internet]. 1999 Dec 14 [cited 2011 Jan 12]. The Tree of Life Web Project. Available from http://tolweb.org/tree?group=Echinodermata &congroup=Metazoa/

6

REVISING

Something that looks like a bad sentence can be the germ of a good one.
LUDWIG WITTGENSTEIN

What a very difficult thing it is to write correctly.
CHARLES DARWIN

No one will ever criticize you for having written too clearly.
NAJ A. KÍNEHCÉP

What Lies Ahead? In This Chapter, You Will Learn

- The importance of completing a first draft early enough to allow at least several days for revision
- How to revise for different sorts of problems in multiple passes through the manuscript
- To always begin by revising for content, and for the logical flow of information and ideas
- How to give constructive criticism on another student's draft, and how to interpret criticism of your own drafts

Much of this book concerns the reading, note taking, thinking, synthesizing, and organizing that permit you to capture your thoughts in a first draft, from which they can't escape. This chapter concerns the revising that must follow: Now, you must examine the first draft critically and diagnose and treat the patient as necessary.

You can't do a thorough job of revising in a single pass. Once you fix the major problems (often relating to the organization of ideas), a whole new set of problems bubbles up to the surface. They then become the next round of major problems that need attention, and so on.

This chapter presents revision as a multistep process. Draft by draft, the product gets better. I typically revise my own writing 4 or 5 times before letting anyone else see it, and several more times after it has been reviewed by others.

All writing benefits from revision. For one thing, the acts of writing and then rereading what you have written typically clarify your thinking. And then, too, even when you finally know precisely what it is you *want* to say, there is the universal difficulty in getting any point across (intact) to a reader. Revising your work improves communication and often leads you to a firmer understanding of what you are writing about . . . and, of course, to a better grade. Here are the steps to follow:

HARDER

- Reorganize your ideas.
- Revise for content—be sure that every sentence says something important and substantive.
- Revise for clarity of expression—if readers have to stop to figure out what you are trying to say or if they misunderstand what you are saying, you have failed.
- Revise for completeness—be specific and complete in making your points.
- Revise for conciseness—say what you need to say in the fewest number of words.
- Revise for flow—link your ideas in logical order, and show that logic to readers.
- Revise for teleology, anthropomorphism, spelling, grammar, and word usage.

EASIER

It is difficult to revise your own work effectively unless you can examine it with a fresh eye. After all, you know what you wanted to say; but without some distance from the work, you can't really tell whether you've actually said it. For this reason, **plan to complete your first draft at least 3 days before the final product is due, to allow time for careful revision**. Reading your paper aloud—and listening to yourself as you read—often reveals weaknesses that you would otherwise miss. It also helps to have one or more fellow students carefully read and comment honestly on your draft; it is always easier to identify writing problems—faulty logic, faulty organization, wordiness, ambiguity, factual errors, spelling and grammatical errors—in the work of others. Forming a peer-editing group is a clear step toward more effective writing and clearer thinking. Be sure to tell readers of your work that you sincerely want constructive criticism, not a pat on the back. Remember, **your goal as a writer is to communicate**, clearly and succinctly, making it as easy as possible for readers to follow your argument. Your goal as a reader of someone else's draft is to help its author do the same. At the end of this chapter, you will find advice on how to be an effective reviewer.

Choose whatever system works best for you, but always revise your papers before submitting them. No matter how sound—or even brilliant—your thoughts and arguments are, it is the manner in which you express them that will determine whether they are understood and appreciated (or, in later life, whether they are even read). With pencil or pen at the ready, the time has come to edit your first draft. Even if you write your first draft on the computer, make at least your first set of revisions on printed copy rather than on-screen; to reorganize effectively, you must see more than one screen of text at a time. Continue revising and editing—printout by printout—until your work is ready for the eyes of the instructor, admissions committee, journal editor, or potential employer. This chapter should help you know when you have arrived at that point.

PREPARING THE DRAFT FOR SURGERY: PLOTTING IDEA MAPS

First drafts are often disorganized messes. Almost always they contain at least a few good ideas, and sometimes they are full of them. But often the ideas are not connected to each other in ways that will seem logical to readers. In reorganizing the material, some ideas can be connected simply by presenting them in a different order. Others can be connected by adding new ideas that will act as bridges between existing ideas. Some ideas cannot be readily connected to the other ideas, however, and do not belong in the same paper, or maybe they belong in different sections of the paper. One way to determine which ideas fall into which category is to sketch out an idea map.*

For the following example, students were asked to write a newspaper article based on a research paper from the primary literature.** Here is what one student's first draft looked like:

1

With their feathers bound by sticky tar, many seabirds could neither fly nor swim. An otter, washed up on the shore, was an unrecoverable black mess. A sudden oil spill killed these animals directly and swiftly. But it doesn't take a catastrophic spill of millions of gallons to cause such devastation.

Crude oil from boat engines, factory effluent, and runoff from city gutters enters the ocean, and takes a toll on some of the least suspected of animals exposed over a period of weeks to the deadly aromatic hydrocarbons that the oil delivers to water.

---

*The approach described here is based on a paper by Flower, L.S., Hayes, J.R. 1977. Problem-solving strategies and the writing process. College English 39: 449–461.

**Stickle, W.B., Rice, S.D., Moles, A. 1984. Bioenergetics and survival of the marine snail *Thais lima* during long-term oil exposure. *Marine Biol.* 80: 281–289.

2 Using a carnivorous marine snail, *Thais lima,* scientists have developed a faster, more sensitive method for assessing toxicity of aromatic hydrocarbons. By measuring growth rates, this method identifies sublethal effects of the toxin, predicting concentrations that will cause eventual death of the animal. A snail with negative growth rate loses more energy than it consumes, so it has none left over to convert to new body tissue. In fact, it starts to burn its own mass to pay the debt.

3 The traditional method used to determine the toxicity of a pollutant is to expose organisms to various concentrations of the pollutant. The concentration that kills half the organisms in a certain amount of time is called the LC-50, which stands for "lethal concentration causing 50% mortality." This has long served as a benchmark for determining toxicity of different compounds in the environment. The lower the LC-50, the more toxic the compound.

While better than short-term assays for predicting effects of pollutants that persist for a long time in the environment, such as aromatic hydrocarbons, long-term assays are costly and time consuming.

4 Scientists typically have exposed invertebrates to aromatic hydrocarbons for 3 days to determine LC-50. In this study by Dr. Bill Stickle and colleagues at Louisiana State University, snails were exposed for up to 28 days.

5 Concentrations of greater than 3,000 parts per billion (ppb) were required to kill half the snails in 3 days; concentrations of only about 800 ppb killed half the snails in 28 days. LC-50s declined with duration of exposure. This means that LC-50s measured after only 3 days give a false picture of animals being more tolerant than they are in the field, where pollutants can persist over much longer periods of time.

6 Growth rates, on the other hand, offer a quicker test for organism health in the presence of a pollutant. 7 In this study, growth rate was determined indirectly by measuring energy intake by the snail (calories from mussel prey) and subtracting energy lost to respiration, feces, and metabolic waste. Growth rates are negative when energy lost is more than energy gained. 8 In *Thais lima*, growth rates were negative when hydrocarbon concentrations exceeded 200 ppb. This concentration is considerably less than the 800 ppb determined to kill half the snails in 28 days. This finding suggests greater sensitivity of the growth rate assay than the LC-50 method to find negative effects of aromatic hydrocarbons.

9 This important study shows how determination of growth rates improves upon traditional measures of toxicity. Use of this technique could refine our knowledge of pollutant effects on marine fauna.

This draft isn't a complete disaster, but it certainly isn't easy to read. The ideas are there, but they aren't well organized. The first 2 paragraphs (1) concern the devastation caused by fuel oils and their components, while the third paragraph (2) discusses ways of measuring toxicity; the beginning of the third paragraph (2) does not follow logically from the previous paragraph. Moreover, the third paragraph raises the issue of developing a faster method for assessing toxicity, while 3 paragraphs later (5), we learn that the "breakthrough" methodology requires 28 days instead of the normal 3 days. That's not faster! Finally, in the next-to-last paragraph (6), we get to the point of the article: Growth rates provide a faster way to judge snail health, by allowing us to predict whether the snails will eventually die. But it isn't a faster way to assess ecosystem health at all; it's a more sensitive method.

The student's seventh paragraph is confusing! The first sentence (6) implies that growth rates *were* determined, the second sentence (7) tells us that growth rates were *not* determined directly, and the fourth sentence (8) implies that growth rates *were* determined after all ("growth rates were negative")! In fact, the researchers did not really monitor growth at all; instead, they measured feeding, respiration, and assimilation rates over a short time and were then able to estimate the extent to which growth rates would be affected. The final sentence of the draft (9) is a giveaway: The vague wording shows that the student has not yet come to grips with what this paper is about. Writing effective newspaper articles is not as easy as it might appear.*

A sentence-by-sentence revision of this piece would be pointless. Massive reorganization is called for. This draft would benefit immensely from idea mapping.

What are the major components of the student's draft?

1. Effects of pollutants
2. LC-50s: What are they?
3. Lethal versus sublethal responses
4. Study used the marine snail, *Thais lima*
5. LC-50 results: 3-day versus 28-day experiments
6. Growth rate measurements: importance; how long does it take to make them?
7. Measured by researchers: respiration rate, feeding rate, assimilation

*See "Science Journalism" under "Writing About Biology PDF's" at http://ase.tufts.edu/biology/faculty/pechenik/ for advice about writing science journalism.

To construct an idea map, scatter the main ideas on a piece of paper, as shown in Figure 9a. Now we need to find a good entry point. I have suggested (Fig. 9b) that we begin by discussing the general issue of pollutant input into marine environments. From there we might discuss lethal versus sublethal responses—but how can we move smoothly between these 2 topics? One possibility would be to note the difficulty of determining the concentration at which any pollutant becomes toxic; this appears as a "bridge" topic in Figure 9b. At first I thought this might lead directly to a discussion of "lethal responses versus sublethal responses," but on further reflection, it seems to lead more logically to a discussion of LC-50s. How to determine toxic pollutant concentrations? Determine the time it takes to cause something to happen in 50% of the animals tested. But that depends on how long the experiment runs. So, we add an arrow from "LC-50s" to the next topic, "short-term versus long-term LC-50 results" (Fig. 9b). From there we can discuss the relative merits of measuring either time to death or some sublethal response, which should be measurable at a lower pollutant concentration (thus increasing the sensitivity of the test). Increased sensitivity is an important point that wasn't in our original list, so I have added it in Figure 9b. Proceeding in this fashion, I have managed to connect all of the ideas presented in Figure 9b to produce a coherent story. It isn't the only way to link the ideas, but it's one way that works. What we end up with is a simple flowchart that can be used to write the next draft of the paper, which should now be pretty easy to write because all of the hard thinking has already been done.

Idea maps can also be used before starting a first draft, but for most people they seem to work best as preparation for the next draft. Once you finally have a coherent story to tell your readers, you can begin to fine-tune the presentation—to make it clear, concise, and fully convincing. The rest of this chapter concerns that fine-tuning process.

REVISING FOR CONTENT

1. **Make sure every sentence says something.** Consider the following opening sentence for an essay on the tolerance of estuarine fish to changes in salinity:

Salinity is a very important factor in marine environments.

What does this sentence say? Does the author really need to tell readers that the ocean is salty? What *is* important about salinity? The sentence is not substantive; it is really just a "running jump," a sentence that may be on the way to something of substance. Let's

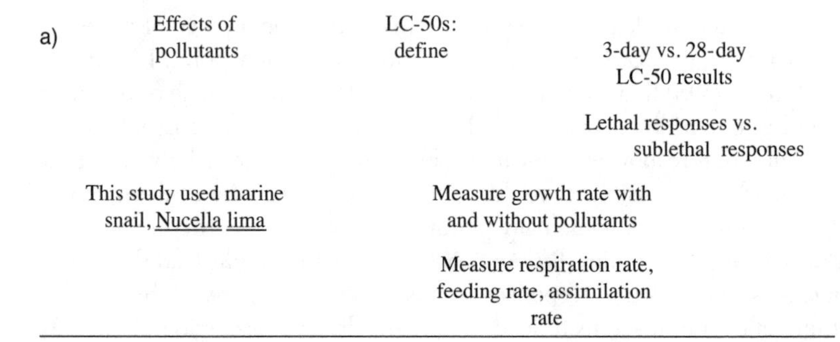

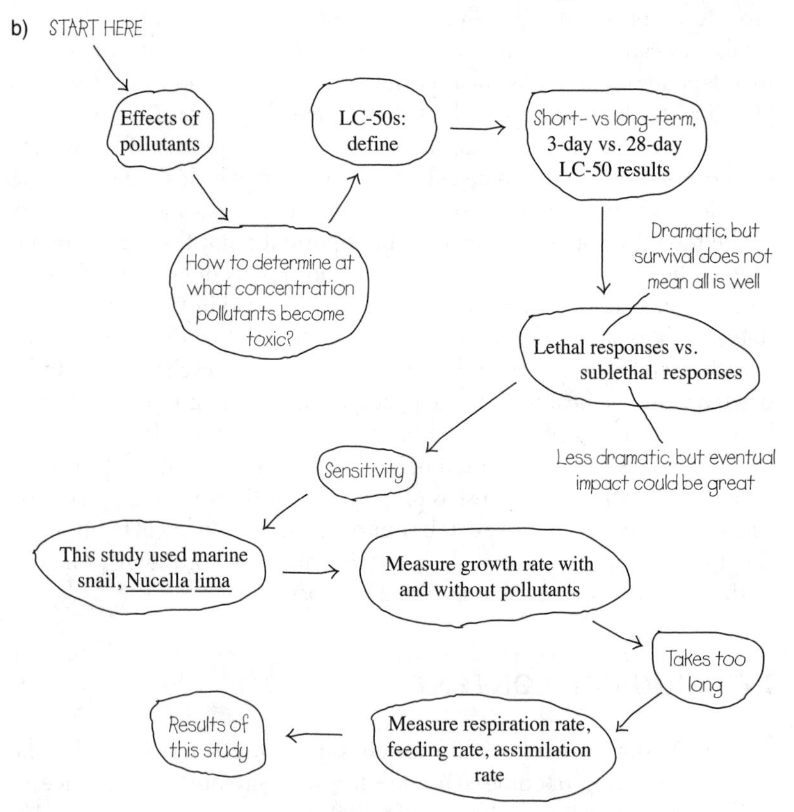

Figure 9. Organizing thoughts by creating idea maps.

a) The main ideas scattered at random, based on a student draft (pp. xx–xx).

b) The ideas connected in one of several possible logical sequences.

delete this sentence and replace it with one that says something worth reading. For example

Estuarine fish may be subjected to enormous changes in salinity within only a few hours.

The author of the revised opening sentence knows where the essay is headed, and so does the reader. The original version got the writer started; the revision process focused the writer's attention on a destination.

Take a careful look at the first sentence of each paragraph that you write for a first draft. You will often find "running jumps," with the substance of the paragraph beginning with your second sentence. Consider the following example:

The damage associated with UV irradiation (280–400 nm wavelengths) on plant and animal populations is well documented. The UV irradiation in sunlight severely damages DNA and other biological molecules in a variety of marine plant and animal species (Gleason and Wellington, 1995; Bingham and Reitzel, 2000; Adams and Shick, 2001).

That first sentence is a running jump if ever I saw one. Cross it out, and get the paragraph off to a much stronger start with a slightly modified form of the second sentence:

The UV irradiation (280–400 nm wavelengths) in sunlight severely damages DNA and other biological molecules in a variety of marine plant and animal species (Gleason and Wellington, 1995; Bingham and Reitzel, 2000; Adams and Shick, 2001).

2. **Include figure references within substantive sentences.** Do not write "The results are shown in Figure 2." Instead, write "Extracts of fungal strains F2 and F14 inhibited the growth of at least 12 bacterial strains (Fig. 2)."
3. **Use the word *relatively* only when making an explicit comparison.** Consider this example:

Many of the animals living near deep-sea hydrothermal vents are relatively large.

The thoughtful reader wonders, "Relative to what?" Either delete the word and replace it with something of substance (e.g., "Animals living near deep-sea hydrothermal vents can exceed lengths of 3 meters") or make a real comparison (e.g., "Many of the animals living near deep-sea hydrothermal vents are large relative to their shallow-water counterparts," or "Some animals living near deep-sea hydrothermal vents are many times larger than their shallow-water counterparts").

4. **Never tell a reader that something is interesting.** Let the reader be the judge. Consider this rather uninformative sentence:

Cell death is a particularly interesting phenomenon.

Is the phenomenon interesting? If so, ask yourself *why* you find it interesting, and then make a statement that will interest the reader. This example could, for instance, be rewritten as follows:

During the development of all animals, certain cells are genetically programmed for an early death.

5. **Be cautious in drawing conclusions, but not overly so.** It is always wise to be careful when interpreting biological data, particularly when you have access to only a few experiments or small data sets. For instance, write "These data suggest that ..." rather than "These data demonstrate that ..." or "These data prove that ..." But don't get carried away, as in the following example:

This suggests the possibility that inductive interactions between cells may be required for the differentiation of nerve tissue.

Here, the author hedges 3 times in 1 sentence, using the words *suggests*, *possibility*, and *may*. Limit yourself to 1 hedge per sentence, as in the following rewrite:

This suggests that inductive interactions are required for the differentiation of nerve tissue.

If you are too unsure of your opinion to write such a sentence, reexamine your opinion.

6. **While revising for content, keep in mind an audience of your peers, not your instructor.** In particular, be sure to define all scientific terms and abbreviations; it is not enough simply to use them properly. Brief definitions will help keep the attention of readers who may not know or may not remember the meaning of some terms, and brief definitions will also demonstrate to your instructor that you know the meaning of the specialized terminology

you are using. **Try to make your writing self-sufficient**; the reader should not have to consult textbooks or other sources to understand what you are saying. For example:

The advantages of outbreeding include reduced exposure of deleterious recessive alleles and increased heterosis, the increased fitness commonly associated with increased heterozygosity.

This would be a better sentence than:

The advantages of outbreeding include reduced exposure of deleterious recessive alleles and increased heterosis.

Note that the author of the better example has cleverly defined the term *heterosis* within the sentence rather than devoting a separate sentence to its definition (see p. 102).

As always, if you write so that you will understand your work years in the future or so that your classmates will understand the work now, your papers and reports will generally have greater impact and will usually earn a higher grade.

REVISING FOR CLARITY

Taming Disobedient Sentences—Sentences That Don't Say What the Author Means

Be sure each sentence says what it's supposed to say; you want the reader's head to be nodding up and down, not side to side. Which way is the reader's head going in the following example?

These methods have different resorption rates and tail shapes.

Do methods have tails? Can methods be resorbed? This sentence certainly fails to communicate what its author had in mind. Here is another sentence that does not reflect the intentions of its author:

From observations made in aquaria, feeding rates of the fish were highest at night.

How many observers do you suppose can fit into an aquarium? Aquaria usually contain fish, not authors; is the author of our example all wet? A revised sentence might read

Feeding rates of fish held in aquaria were highest at night.

Some biologists are clearly more dedicated to their research than most of us are:

Ferguson (1963) examined autoradiographs of sea star digestive tissue after being fed radioactive clams.

Perhaps we should feed the clams not to Ferguson but to the sea stars?

Ferguson (1963) fed radioactive clams to sea stars and then examined autoradiographs of the sea star digestive tissue.

In the preceding example, **note the advantages of summarizing a study in the order in which steps were undertaken**. Grammatical difficulties typically vanish, and the sentence automatically becomes clearer.

Sometimes sentences are confusing because the author tries to stuff too much into them, as in the following example:

The Coomassie blue stain, a nonspecific dye that binds to all proteins, was used to show all the proteins in the samples allowing analysis of induction of the appearance of a band after induction that is not found in the uninduced sample.

Whew! The goal of the experiment was to induce certain bacteria to express a cloned gene. The bacteria were then homogenized and their proteins separated on an electrophoretic gel. The gel was then stained with Coomassie blue to visualize the proteins present. If a new protein (the induced protein) was present, it would show up as a new, separate band. Now that's just too much work for any 1 sentence to accomplish!

Confusing sentences also inevitably arise when 3 or more nouns are lined up in a row. Consider this example:

Sleep study results show that tryptophan significantly decreases the time needed to fall asleep (Miller and Brown, 2007).

At the first reading, the reader probably expects "results" to be a verb, but instead, it is a noun, preceded by 2 other nouns. The reader must stop and decode the sentence. Ah! The author is discussing the results of studies of people sleeping. We can rewrite the sentence to make this much clearer:

Recent studies show that tryptophan decreases the time needed for people to fall asleep (Miller and Brown, 2007).

In revising your work, **think twice before leaving more than 2 nouns together**: 2 is company; 3 is a crowd.

Here are 2 additional examples of unclear writing:

This determination was based on mannitol's relative toxicity to sodium chloride.

> The surface area of mammalian small intestines is 3 to 7 times greater than reptiles.

With the first example, how can one chemical be toxic to another chemical? The author is probably trying to tell us that 2 chemicals differ in their toxicity to some organisms or cell types. With the second example, one wonders how an intestinal surface area can be greater than a reptile; again, the author is not making the comparison he or she intended.

Readers should never have to guess what the proper comparisons are; **the less you make your readers work, the more they will appreciate your writing**. In any event, never invoke the "You know what I mean" defense. If a student writes, "A normal human fetus has 46 chromosomes," how can I assume the student understands that each *cell* of the fetus has 46 chromosomes? **It is your job to inform the reader; it is never the reader's job to guess what you are trying to say**.

Of course, it doesn't help that confusing sentences surround us in our everyday lives. Consider this example taken from a local newspaper:

> Offer void where prohibited by law, or while supplies last.

Apparently, no one can ever take advantage of this offer; the offer is either prohibited by law or, if permitted by law, is void while supplies last. Supplies can never run out since the advertiser seems unwilling to fill your order as long as the items are available. If stocks become depleted, perhaps by eventual disintegration of the product, the advertiser could then honor your request, but the company would no longer have anything to send you!

Then there are these gems:

> It is illegal for school workers to say that children should take a psychiatric drug because they are not doctors.

> Our nuclear reactors are as safe as they can possibly be. And we are constantly working to make them safer.

With practice, you can find similarly confusing or absurd sentences almost anywhere you look; if they appear in your own writing, revise them. Make each sentence state its case unambiguously. Here is a sentence that does not do so:

> Sea stars prey on a wide range of intertidal animals, depending on their size.

Is the author talking about the size of the sea stars that are preying or about the size of the intertidal animals that are preyed upon? Don't be embarrassed at finding sentences like this one in early drafts of your papers and reports. Be embarrassed only when you don't edit them out of your final draft.

The Dangers of *It*

Frequent use of the pronouns *it*, *they*, *these*, *their*, *this*, and *them* in your writing should sound an alarm: Probable ambiguity ahead. Consider the following example of the trouble *it* can cause

The body is covered by a cuticle, but it is unwaxed.

Which is unwaxed: the body or the cuticle? *It* makes the second part of the following sentence equally ambiguous:

The chemical signal must then be transported to the specific target tissue, but it is effective only if it possesses appropriate receptors.

Are these receptors needed by the chemical signal or by the target tissue? I'm confused.

In the next example, *these* causes similar problems for readers:

Antigens encounter lymphocytes in the spleen, tonsils, and other secondary lymphoid organs. These then proliferate and differentiate into fully mature, antigen-specific effector cells.

Presumably the lymphocytes, not the tonsils, are proliferating, although the author has certainly not made this clear. The problem is easily repaired by beginning the second sentence with "The lymphocytes ..." In the next example, *their* is guilty of a similar offense:

Like fanworms and earthworms, leeches have proven very useful to neurophysiologists. Their neurons are few and large, making them particularly easy to study with electrodes.

Most readers are likely surprised to learn that neurophysiologists have so few neurons and are so easy to study.

Now let us consider the difficulties *they* can cause

Tropical countries are home to both venomous and nonvenomous snakes. They kill their prey by constriction or by biting and swallowing them.

How much clearer the last sentence could become by replacing *they* with a few words of substance and by deleting *them* entirely:

Tropical countries are home to both venomous and nonvenomous snakes. The nonvenomous snakes kill their prey by constriction or by biting and swallowing.

And have you ever met researchers who glowed in the dark? One student apparently has

Harper and Case (1999) found that the plainfish midshipman, *Porichthythys notatus*, experienced twice the rate of predation when they were not luminous.

If *they* have their way, the reader must guess who was glowing. Realizing that the sentence is in difficulty, we revise

Harper and Case (1999) found that luminous plainfish midshipman (*Porichthythys notatus*) were twice as likely to be eaten as nonluminous specimens.

Finally, look what can happen when a variety of these pronouns are scattered throughout a sentence:

Although *they* both saw the same things in *their* observations of embryonic development, *they* had different theories about how *this* came about.

A patient reader of the whole essay could probably figure out this sentence, eventually, but its author has certainly violated one of our key rules, "**Never make the reader back up**," in a most extreme fashion.

In short, when revising your work, read it carefully and with skepticism, checking that you have said exactly what you mean. **Never make the reader guess what you have in mind.** Never give the reader cause to wonder whether, in fact, you *have* anything in mind. Everything you write must make sense—to you and to the reader. As you read each sentence that you have written, think these: What does this sentence really say? What did I mean it to say? Make each sentence work on your behalf, leading the reader easily from fact to fact, from thought to thought.

Please note that you need not be a grammarian to write correctly and clearly. With a little practice, especially if you read your work aloud, you can quickly learn to recognize a sentence in difficulty and sense how to fix it without even knowing the name of the grammatical rule that was violated.

Problems with *And*

The word *and* can sometimes obscure connections between ideas. Consider this example:

Thorson (1950) suggested that sea star recruitment is often low despite the release of tens of millions of eggs because the probability of eggs encountering sperm in the plankton is low, and that most eggs are never fertilized.

The connection between the first and last parts of that sentence is unclear. A simple revision, in which *and* is replaced by *so*, clarifies the connection:

Thorson (1950) suggested that sea star recruitment is often low despite the release of tens of millions of eggs because the probability of eggs encountering sperm in the plankton is low, so that most eggs are never fertilized.

The word *and* can also weaken a sentence by linking ideas that would be better left on their own, as in this example:

In two similar habitats, one with fish and one without fish, there was a negative correlation between the numbers of frogs and the numbers of fish and researchers believe that the accidental introduction of the fish is responsible for 60% of the variation in the distribution of frogs in these two regions (Knapp and Matthews, 2000).

Try rereading the above sentence after replacing the *and* with a period and breaking it into 2 sentences: Now the second idea gets the attention it deserves.

Be sure that *and* is not being used to link weakly related or unrelated ideas.

Headache by Acronym

Overuse of acronyms can drive away potentially interested readers. Some acronyms, such as DNA, are widely known and are fine to use. But consider this sentence:

Within 24 hours of induction by 5-HT, most cells in the AG showed signs of PCD.

Why would any author do something like this to a potentially interested reader? I would rewrite that sentence as follows:

Within 24 hours of induction by 5-HT (serotonin), most cells in the apical ganglion showed signs of programmed cell death.

Avoid acronyms.

REVISING FOR COMPLETENESS

Make sure each thought is complete. **Be specific** in making assertions. The following statement is much too vague:

Many insect species have been described.

How many is "many"? After editing, the sentence might read

Nearly 1 million insect species have been described.

Similarly, the sentence

More caterpillars chose diet *A* than diet *B* when given a choice of the 2 diets (Fig. 2).

would benefit from the following alteration:

Nearly 5 times as many caterpillars chose diet *A* than diet *B* when given a choice between the 2 diets (Fig. 2).

Here is another kind of incompleteness:

If diffusion was entirely responsible for glucose transport, then this would not have occurred.

This rears its ugly head again; the author avoids the responsibility of drawing a clear conclusion and forces the reader to back up and attempt to summarize the findings. Even the beginning of the sentence is unnecessarily vague because, it turns out, the discussion is concerned only with glucose transport in intestinal tissue. Try to make your sentences tell a more detailed story, as in this revision:

If diffusion was entirely responsible for glucose transport into cells of the intestinal epithelium, transport would have continued when I added the inhibitors.

In the same way, "Cells exposed to copper chloride divided at normal rates" is a substantial improvement over "The copper chloride treatment was not affected."

Be especially careful to revise for completeness whenever you find that you have written *etc.*, an abbreviation for the Latin term *et cetera*, meaning "and others" or "and so forth." In writing a first draft, use *etc.* freely when you'd rather not interrupt the flow of your thoughts by thinking about exactly what "other things" you have in mind. When revising, however, **replace each *etc.* with words of substance**; in scientific writing, an *etc.* makes the reader suspect fuzzy thinking. Ask yourself, "What, exactly, *do* I have in mind here?" If you come up with additional items for your list, add them. If you find that you have nothing to add, simply replace the *etc.* with a period, and you will have produced a shorter, clearer sentence.

Consider the following sentence and its 2 improvements:

ORIGINAL VERSION

Plant growth is influenced by a variety of environmental factors, such as light intensity, nutrient availability, etc.

REVISION 1

Plant growth is influenced by a variety of environmental factors, such as light intensity, day length, nutrient availability, and temperature.

REVISION 2

Plant growth is influenced by such environmental factors as light intensity, day length, nutrient availability, and temperature.

In the original version, the author dodged the responsibility of clear, informative writing, forcing the reader to determine what is meant by *etc.* The sentence, although grammatically correct, is incomplete, waiting for the reader to fill in the missing information. The reader may justifiably wonder whether the writer knows what other factors affect plant growth. Both revised versions clearly indicate what the author had in mind. **Revising for completeness often requires you to return to your notes or to the sources upon which your notes are based.**

REVISING FOR CONCISENESS

Make every word count. Omitting unnecessary words will make your thoughts clearer and more convincing. I have already talked about entire sentences that are really nothing more than running jumps, particularly at the beginning of paragraphs (p. 85). **Often you can find running jumps at the start of sentences**, too. In particular, such phrases as "It should be noted that," "It is interesting to note that," "Evidence has shown that," "It has been documented that," "Analysis of the data indicated that," and "The fact of the matter is that" are common in first drafts but should be ruthlessly eliminated in preparing the second.

Consider an example:

The data indicate that as hermit crabs gain in size and weight, they tend to occupy larger shells (Fig. 3).

To revise, we merely eradicate those first 4 words:

As hermit crabs gain in size and weight, they tend to occupy larger shells (Fig. 3).

Here's an example of an even longer running jump:

Evidence provided by Clarke and Faulkes (1999) has shown that reproductively mature female mole rats (*Heterocephalus*

glaber) preferentially mate with unrelated males, reducing the frequency of inbreeding.

That sentence doesn't really become substantive until we get to the mature females, so let's remove the running jump to end up with:

Reproductively mature female mole rats (*Heterocephalus glaber*) preferentially mated with unrelated males, reducing the frequency of inbreeding (Clarke and Faulkes 1999).

Running jumps often find their way into sentences presenting the results of statistical analyses, as in this example:

Chi-square analysis of the samples collected at Fanghorn Wood, Middle Earth, indicated that there was no significant difference in the proportion of occupied intact pine cones compared to the proportion of occupied damaged pine cones ($\chi^2 = 0.26$; *d.f.* = 1; $p = 0.61$).

Everything preceding and including the words "indicated that" constitutes a running jump. Let readers dive right in:

At Fanghorn Wood, Middle Earth, the proportion of pine cones occupied by weevils was not significantly affected by whether the pine cones were damaged ($\chi^2 = 0.26$; *d.f.* = 1; $p = 0.61$).

I have discussed writing about statistics more fully in Chapter 4.

Verbal excess can also take less conspicuous forms. How might you shorten this next sentence?

Dr. Smith's research investigated the effect of pesticides on the reproductive biology of birds.

Who did the work: Dr. Smith or her research? A reasonable revision would

Dr. Smith investigated the effect of pesticides on the reproductive biology of birds.

We have eliminated one word, and the sentence has not suffered a bit. Working on the sentence further, we can replace "the reproductive biology of birds" with "avian reproduction," achieving a net reduction of 3 more words:

Dr. Smith investigated the effect of pesticides on avian reproduction.

The next example requires similar attention:

It was found that the shell lengths of live snails tended to be larger for individuals collected closer to the low tide mark (Fig. 1).

A good editor would eliminate the first phrase of that sentence and prune further from there. In particular, what does the author mean by "tended to be larger"? Here are 2 improved versions of the sentence:

Live snails collected near the low tide mark had greater average shell lengths (Fig. 1).

Snails found closer to the low tide mark typically had larger shells (Fig. 1).

These and most other wordy sentences suffer from one or several of 4 major ailments and can be brought to robust health by obeying the following Five Commandments of Concise Writing.

First Commandment: Eliminate Unnecessary Prepositions

Consider this example:

The results indicated a role of hemal tissue in moving nutritive substances to the gonads of the animal.

Any sentence containing such a long string of prepositional phrases—"of … tissue," "in moving … substances," "to the gonads," "of the animal"—is automatically a candidate for the editor's operating table. This sentence actually contains a simple thought, buried amid a clutter of unnecessary words. After surgery, the thought emerges clearly:

The results indicated that hemal tissue moved nutrients to the animal's gonads.

Here is another example:

The cells respond to foreign proteins by rapidly dividing and starting to produce antibodies reactive to the protein groups that induced their production.

The reader's head spins, an effect avoided by the following more concise incarnation of the same sentence:

In the presence of foreign proteins, the cells divide rapidly and produce antibodies against those proteins.

By eliminating prepositions, "Karlson arrives at the conclusion that …" becomes "Karlson concludes that …"; "Grazing may constitute a benefit to …" becomes "Grazing may benefit …"; and "These data appear to be in support of the hypothesis that … " becomes "These data appear to support the hypothesis that ..." or perhaps "These data support the hypothesis that ..."

Second Commandment: Avoid Weak Verbs

Formal scientific writing is often confusing—and boring—because the individual sentences contain no real action; commonly, the colorless verb *to be* is used where a more vivid verb would be more effective, as in this example:

The fidelity of DNA replication is dependent on the fact that DNA is a double-stranded polymer held together by weak chemical interactions between the nucleotides on opposite DNA strands.

This patient suffers from **Wimpy Verb Syndrome**, with a slight touch of Excess Prepositional Phrase. There is *potential* action in the sentence, but it is sound asleep in the verb "is dependent." Converting to the stronger verb "depends," we read:

The fidelity of DNA replication depends on the fact that DNA is a double-stranded polymer ...

But why stop there? Let's eliminate some clutter ("on the fact that") and another weak verb ("is"):

The fidelity of DNA replication depends on DNA being a double-stranded polymer ...

Along the same lines, can you find a potentially stronger verb in this next sentence?

Activation of the immune response may be a trigger for disease progression (Bernheim, 1980).

Why not replace "may be a trigger" [yawn] with "may trigger" [more exciting]? Then the sentence becomes:

Activation of the immune response may trigger disease progression (Bernheim, 1980).

Similarly:

Plant vascular tissues function in the transport of food through xylem and phloem.

can be enlivened by converting the phrase "function in the transport of" to the more vigorous verb "transport":

Plant vascular tissues transport food through xylem and phloem.

Note that by choosing a stronger verb, we have also eliminated 2 prepositional phrases ("in the transport" and "of food"). Step by step, the sentence becomes shorter and clearer. But, **as often happens during revision, fixing one problem reveals an additional problem**, in this case a fundamental structural weakness that makes the reader wonder whether the student understands the relationship between "plant vascular tissues" and "xylem and phloem." Revising now for content, we might rewrite the sentence as:

Plant vascular tissues (the xylem and phloem) transport nutrients throughout the plant.

or

Plants transport nutrients through their vascular tissues, the xylem and phloem.

Third Commandment: Do Not Overuse the Passive Voice

Although passive voice has its uses, it is often a great enemy of concise writing, in part because the associated verbs are weak. If the subject ("Rats and mice," in the following example) is on the receiving end of the action, the voice is passive:

Rats and mice were experimented on by him.

If, on the other hand, the subject of a sentence ("He," in the following example) is on the delivering end of the action, the voice is said to be active:

He experimented with rats and mice.

Note that the active sentence contains only 6 words, whereas its passive counterpart contains 8. In addition to creating excessively wordy sentences, the passive voice often makes the active agent anonymous, and a weaker, sometimes ambiguous sentence may result:

Once every month for 2 years, mussels were collected from 5 intertidal sites in Barnstable County, MA.

Who should the reader contact if there is a question about where the mussels were collected? Were the mussels collected by the writer, by fellow students, by an instructor, or by a private company? Eliminating the passive voice clarifies the procedure:

Once every month for 2 years, members of the class collected mussels from 5 intertidal sites in Barnstable County, MA.

Similarly, "It was found that" becomes "I found," or "We found," or perhaps, "Karlson (1996) found." Whenever it is important, or at least useful, that the reader know who the agent of the action is, and whenever the passive voice makes a sentence unnecessarily wordy, use the active voice:

Passive: Little is known of the nutritional requirements of these animals.

Active: We know little about the nutritional requirements of these animals.

Passive: The results were interpreted as indicative of . . .

Active: The results indicated . . .

Passive: In the present study, the food value of 7 diets was compared, and the chemical composition of each diet was correlated with its nutritional value.

Active: In this study, I compared the food value of 7 diets and correlated the chemical composition of each diet with its nutritional value.

Note in this last example that it is perfectly acceptable to use the pronoun *I* in scientific writing; switching to the active voice expresses thoughts more forcibly and clearly and often eliminates unnecessary words.

Fourth Commandment: Make the Organism the Agent of the Action

Consider this example:

Studies on the rat show that the activity levels vary predictably during the day (Hatter, 1976).

This is not a terrible sentence, but it can be improved by moving the action from the studies ("Studies . . . show") to the organism involved, the rat:

Rats vary their activity levels predictably during the day (Hatter, 1976).

The revised sentence is shorter, clearer, and more interesting because now an organism is *doing* something. Along the way, a prepositional phrase ("on the rat") has vanished. Alternatively, one could include the researcher in the action:

Hatter (1976) showed that rats vary their activity levels predictably during the day.

Similarly, redirecting the action transforms:

> The reaction rate increased as pH was increased from 6.0 to about 8.0 and then declined between a pH of 8.5 and 9.0 (Fig. 1).

to

> Trypsin was maximally effective at pHs between about 8.0 and 8.5 (Fig. 1).

Note that in the original version, the author redrew a graph in words: We can easily picture the author staring at the graph and its axes while writing. In the revised version, the author makes the enzyme the agent of the action, and the message comes through much more clearly.

Be a person of few words; your readers will be grateful.

Fifth Commandment: Incorporate Definitions into Your Sentences

If you find yourself writing something like "Kairomones are defined as ..." you're on your way to writing a wordy and boring sentence. Instead, try to incorporate definitions into substantive sentences that move your presentation forward. For example, you might write something like this:

> Chemicals (known as "kairomones") that are released by predators can sometimes induce morphological changes in their crustacean prey.

or

> Chemicals released into the water by predators can sometimes induce morphological changes in their prey. Among some crustacean prey species, for example, such "kairomones" can induce 50% increases in ...

REVISING FOR FLOW

A strong paragraph—indeed, a strong paper—takes the reader smoothly and inevitably from a point upstream to one downstream. Link your sentences and paragraphs using appropriate transitions so that the reader moves effortlessly and inevitably, logically and unambiguously, from one thought to the next. Minimize turbulence. Remind readers of what has come before, and help them anticipate what is coming next. Consider the following example:

> Since aquatic organisms are in no danger of drying out, gas exchange can occur across the general body surface. The body walls of aquatic invertebrates are generally thin and water

permeable. Terrestrial species that rely on simple diffusion of gases through unspecialized body surfaces must have some means of maintaining a moist body surface or must have an impermeable outer body surface to prevent dehydration; gas exchange must occur through specialized, internal respiratory structures.

This example gives the reader a choppy ride indeed, and it cries out for careful revision, not of the ideas themselves but of the way they are presented. In the following revision, note the effect of 2 important transitional expressions, *thus* and *in contrast to*. The first connects 2 thoughts; the second warns the reader of an approaching shift in direction:

Since aquatic organisms are in no danger of drying out, gas exchange can occur across the general body surface. Thus, the body walls of aquatic invertebrates are generally thin and water permeable, facilitating such gas exchange. In contrast to the simplicity of gas exchange mechanisms among aquatic species, terrestrial species that rely on simple diffusion of gases through unspecialized body surfaces must either have some means of maintaining a moist body surface or must have an impermeable outer body covering that prevents dehydration. If the outer body wall is impermeable to water and gases, respiratory structures must be specialized and internal.

In the first draft, the reader must struggle to find the connection between sentences. In the revised version, the writer has assisted the reader by connecting the thoughts, resulting in a more coherent paragraph.

Here is one more example of a stagnating paragraph that carries its readers nowhere:

The energy needs of a resting sea otter are 3 times those of terrestrial animals of comparable size. The sea otter must eat about 25% of its body weight daily. Sea otters feed at night as well as during the day.

Revising for improved flow, or coherence, produces the following paragraph. Note that the writer has introduced no new ideas. The additions, here underlined, are simply clarifications that make the connections between each point explicit:

The energy needs of a resting sea otter are 3 times those of terrestrial animals of comparable size. <u>To support such a high metabolic rate</u>, the sea otter must eat about 25% of its body weight daily. <u>Moreover</u>, sea otters feed continually, at night as well as during the day.

The following transitional words and phrases are especially useful in linking thoughts to improve flow: *in contrast, however, although, for example, thus, whereas, even so, nevertheless, moreover, despite, in addition to.* The use of such words can also help readers see connections between adjacent paragraphs, as in the following example (connecting words are underlined):

> Decreased fecundity due to inbreeding depression is well documented in plants. [The student then gives several examples] . . . and in *Trillium erectum*, self-pollinated plants produced 71% fewer seeds than outcrossed plants (Irwin, 2001).
>
> Similar effects of inbreeding on fecundity have been reported in a number of bird species. For example, . . .

Repetition and summary are highly effective ways to link thoughts. For instance, repetition was used to connect the first 2 sentences of the revised example about sea otters: "To support such a high metabolic rate" essentially repeats, in summary form, the information content of the first sentence. Repetition was also used to link the paragraphs about inbreeding that you just read: In reminding readers of what has come before, the author consolidates his or her position and then moves on. Use these and similar transitions to move the reader smoothly from the beginning of your paper to the end. Be certain that each sentence—and each paragraph—sets the stage for the one that follows, and that each sentence—and each paragraph—builds on the one that came before.

A Short Exercise in Establishing Coherence

Your sentences should lead so logically and smoothly from one thought to the next that should the individual sentences of a paragraph become scattered by heavy winds, someone who collects all of those individual sentences should be able to reassemble the original paragraph. For example, I have deliberately disassembled a paragraph* into the following isolated sentences:

a. It is becoming clear, however, that although wave propagation is a common feature of activation, there are both subtle and significant differences in this response when comparing eggs from different species.

*From Lee, K.W., Webb, S.E., Miller, A.L. 1999. A wave of free cytosolic calcium traverses zebrafish eggs on activation. *Devel. Biol.* 214: 168–180.

b. It appears that all vertebrate, invertebrate, and perhaps even some plant eggs are activated by the generation of calcium transients in their cytoplasm (Roberts *et al.*, 1994; Lawrence *et al.*, 1997).
c. In contrast, activation triggers a series of repetitive calcium waves or oscillations in annelids (Stricker, 1996), ascidians (Albrieux *et al.*, 1997), and mammals (Kline and Kline, 1992), including humans (Homa and Swann, 1994; Tesarik and Testart, 1994).
d. For example, in fish (Gilkey *et al.*, 1978), echinoderms (Stricker *et al.*, 1992), and frogs (Busa and Nuccitelli, 1985; Kubota *et al.*, 1987), a single calcium wave is propagated across the activating egg.
e. In most cases these transients take the form of propagating calcium waves (Jaffe, 1985; Epel, 1990; Whitaker and Swamm, 1993), which appear to be essential for activating the eggs.

Try reconstructing the original paragraph. Clearly, sentence "a" can't be the opening sentence of the paragraph. Why not? What sentence most likely precedes sentence "a"? What sentence most likely follows sentence "a"? What words provide the clues that allow you to answer these questions? Which sentence provides the most general statement of the problem? I nominate that sentence as our best candidate for Opening Sentence of the Paragraph. The correct sentence order is shown at the end of this chapter (p. 122). Try to make your own paragraphs as easy to reconstruct, in part by using the tricks of repetition and summary and by using appropriate transitional words and phrases.

Improving Flow Using Punctuation

Judicious use of the semicolon can also ease the reader's journey. In particular, when the second sentence of a pair explains or clarifies something contained in the first, you may wish to combine the 2 sentences into one with a semicolon. Consider the following 2 sentences:

This enlarged and modified bone, with its associated muscles, serves as a useful adaptation for the panda. With its "thumb," the panda can easily strip the bamboo on which it feeds.

The reader probably has to pause to consider the connection between the 2 sentences. Using a semicolon, the passage would read

This enlarged and modified bone, with its associated muscles, serves as a useful adaptation for the panda; with its "thumb," the panda can easily strip the bamboo on which it feeds.

The semicolon links the 2 sentences and eliminates an obstruction in the reader's path.

Similarly, a semicolon provides an effective connection between thoughts in the following 2 examples:

Recently we demonstrated the rapid germination of radish seeds; nearly 80% of the seeds germinated within 3 days of planting.

Recombinant DNA technology enables large-scale production of particular gene products; specific genes are transferred to rapidly dividing host organisms (yeast or bacteria), which then transcribe and translate the introduced genetic templates.

REVISING FOR TELEOLOGY AND ANTHROPOMORPHISM

Remember, **most organisms do not act or evolve with intent** (p. 10). Consider the following examples of teleological writing, and learn to avoid it in your own work:

> Barnacles cannot move from place to place and therefore had to evolve a specialized food-collecting apparatus in order to survive.
>
> Squid and most other cephalopods lost their external shells in order to swim faster, and so better compete with fish.
>
> Many animals use antipredator behaviors to increase their chance of survival.

Revise all teleology out of your writing. Don't have nonhuman animals thinking and planning.

Also beware of anthropomorphizing, in which you give human characteristics to nonhuman entities, as in this example:

> The existence of sage in the harsh climate of the American plains results from Nature's timeless experimentation.

Again, this conveys a rather fuzzy picture about how natural selection operates. The author would be on firmer ground by writing something like this:

> Sage is one of the few plants capable of withstanding the harsh, dry climate of the American plains.

REVISING FOR SPELLING ERRORS

Misspellings convey the impression of carelessness, laziness, or perhaps even stupidity. These are not advisable images to present to instructors, prospective employers, or the admissions officers of graduate or professional programs. Spell-check computer programs can save you from misspelling many nontechnical words, but they won't catch such spelling errors as *is* versus *if*, or *nothing* versus *noting*, and are unlikely to be of much help in screening technical terms for you. Use the computer for a first pass, but use your own eyes for the second.

It helps to keep a list of words that you find yourself using often and consistently misspelling. *Desiccation* and *argument* were on my list for quite some time; *proceed* and *precede* are still on it. When in doubt, use a dictionary. And if you add technical terms to your computer program's dictionary, be careful to enter the correct spellings.

A few peculiarities of the English language are worth pointing out:

1. *Mucus* is a noun; as an adjective, the same slime becomes *mucous*. Thus, many marine animals produce mucus, and many marine animals produce mucous trails.
2. *Seawater* is always a single word. *Fresh water*, however, is usually 2 words as a noun and 1 word as an adjective. Thus, freshwater animals live in fresh water. The Council of Science Editors no longer insists on this usage, however, and different publishers are setting their own rules.
3. *Species* is both singular and plural: 1 species, 2 species. But the plural of *genus* is *genera*: 1 genus, 2 genera.
4. The plural forms of *alga*, *bacterium*, and *hypothesis* are *algae*, *bacteria*, and *hypotheses*.
5. When writing about insect larvae, *worm* is never written as a separate word, because insect larvae are not true worms (i.e., they are not annelids). One studies *silkworms*, for example, not *silk worms*. Similarly, for 2-part insect names, the second part (e.g., *fly* or *bug*) is never written as a separate word when it is not correct systematically, but otherwise is written as a separate word. We write, for example, about *butterflies* (which do not belong to the order Diptera, containing the true flies) and about *house flies* (which *are* true flies) and *bed bugs* (which *are* true bugs, members of the order Hemiptera).

See also page 110 for the distinction between *effect* and *affect*. And **don't forget to underline or italicize scientific names**: *Littorina littorea* (the periwinkle snail), *Chrysemys picta* (the eastern painted turtle), *Taraxacum officinale* (the common dandelion), *Caenorhabditis elegans* (a nematode worm, the first animal to have its entire genome sequenced), *Homo sapiens* (the only animal that writes laboratory reports).

REVISING FOR GRAMMAR AND PROPER WORD USAGE

Appendix C lists a number of books and websites that include excellent sections on grammar and proper word usage. While on the lookout for sentence fragments, run-on sentences, faulty use of commas, faulty parallelism, incorrect agreement between subjects and verbs, and other grammatical blunders, you should also be on the lookout for violations of 11 especially troublesome rules of usage when revising your work.

1. *between* and *among*. *Between* (from *by twain*) usually refers to only 2 things:

    The 20 caterpillars were randomly distributed between the 2 dishes.

    Among usually refers to more than 2 things:

    The 20 caterpillars were randomly distributed among the 8 dishes.

2. *which* and *that*. Most of your *which*s should be *that*s:

    This fish, which lives at depths up to 1,000 m, experiences up to 101 atmospheres of pressure.

    A fish that lives at a depth of 1,000 m is exposed to 101 atmospheres of pressure.

    In the first example, *which* introduces a nondefining, or nonrestrictive, clause. The introduced phrase is, in effect, an aside, adding extra information about the fish in question; the sentence would survive without it. On the other hand, the *that* in the second example introduces a defining, or restrictive, clause; we are being told about a particular fish, or type of fish, one that lives at a depth of 1,000 m.

    Improper use of *that* and *which* can occasionally lead to ambiguity or falsehood. Consider the following sentence about the production of proteins from messenger RNA (mRNA) transcripts:

    This difference in protein production is due to different amounts of mRNA that translate and produce each particular protein.

    Here, *that* correctly introduces a restrictive clause. Which mRNA molecules? The ones coding for these particular proteins.

The writer is telling us that proteins are produced in proportion to the number of mRNA molecules coding for them within the cell. Replacing *that* with *which* drastically changes the meaning of the sentence:

> The difference in protein production is due to different amounts of mRNA, which translates and produces each particular protein.

The sentence has lost clarity because *which* now introduces a nondefining clause that should be explaining only what mRNA does in general. In the following sentence, using the word *which* conveys information that is actually wrong:

> In squid and other cephalopods, which lack external shells, locomotion is accomplished by contracting the muscular mantle.

Here, the writer asserts that no cephalopods have external shells, which is not the case; some species *do* have external shells. The correct word is *that*:

> In squid and other cephalopods that lack external shells . . .

Now the writer correctly refers specifically to those cephalopods without external shells.

As in the examples given, *which* is commonly preceded by a comma. When deciding between *which* and *that* in your own writing, read your sentence aloud. If the word doesn't need a comma before it for the sentence to make sense, the correct word is probably *that.* If you hear a pause when you read, signifying the need for a comma, the correct word is probably *which.*

3. *its* and *it's. It's* is always an abbreviated form of *it is.* If *it is* does not belong in your sentence, use the possessive pronoun *its*:

> When treated with the chemical, the protozoan lost its cilia and died.

> It's clear that the loss of cilia was caused by treatment with the chemical.

While we're at it, let's revise that last sentence to eliminate the passive voice:

> It's clear that treatment with the chemical caused the loss of cilia.

In general, contractions are not welcome in formal scientific writing. Thus, you can avoid the problem entirely by writing *it is* when appropriate:

> It is clear that treatment with the chemical caused the loss of cilia.

4. *effect* and *affect*. *Effect* as a noun means a "result" or "outcome":

   > What is the effect of fuel oil on the feeding behavior of sea birds?

   Effect as a verb means "to bring about":

   > What changes in feeding behavior will fuel oil effect in sea birds?

   Affect as a verb means "to influence" or "to produce an effect upon":

   > How will the fuel oil affect the feeding behavior of sea birds?

   Used as a verb, *effect* can, indeed, be replaced in the preceding example by *bring about*, but not by *influence*; and *affect* can, indeed, be replaced by *influence*, but not by *bring about*. Even so, memorizing the definitions of the 2 words may be of little help in deciding which word to use in your own writing because, as verbs, *affect* and *effect* are so similar in meaning. You may be more successful in choosing the correct word by memorizing each of the above examples and then comparing the memorized examples with your own sentences.
5. *rate*. "Rates" have units of "something per time": moles of substrate degraded per minute, numbers of centimeters (cm) moved per second, numbers of births per year, and so forth. If you are writing about something that does *not* have units of "per time," then do not use the word *rate*.
6. *i.e.* and *e.g.* These 2 abbreviations are not interchangeable. The abbreviation *i.e.* is an abbreviation for *id est*, which, in Latin, means "that is" or "that is to say." For example:

   > Data on sex determination suggest that this species has only two sexual genotypes, i.e., female (XX) and male (XY).

   > The embryos were undifferentiated at this stage of development; i.e., they lacked external cilia and the gut had not yet formed.

In contrast, *e.g.* stands for *exempli gratia*, which means "for example." I will give 2 examples of its use:

> During the precompetent period of development, the larvae cannot be induced to metamorphose (e.g., Crisp, 1974; Bonar, 1978; Chia, 1978; Pires, 2000).

> However, the larvae of several butterfly species (e.g., *Papilio demodocus* Esper, *P. eurymedon*, and *Pieris napi*) are able to feed and grow on plants that the adults never lay eggs on.

In the first case, the writer uses *e.g.* to indicate that what follows is only a partial listing of references supporting the statement: "for example, see these references," in other words. In the second case, the writer uses it to indicate only a partial list of butterfly species that do not lay eggs on all suitable plants.

7. *However, therefore,* and *moreover.* These words are often incorrectly used as conjunctions, as in the following examples:

> The brain of a toothed whale is larger than the human brain, however the ratio of brain to body weight is greater in humans.

> The resistance of mosquito fish (*Gambusia affinis*) to the pesticide DDT persisted into the next generation bred in the laboratory, therefore the resistance was probably genetically based.

> Protein synthesis in frog eggs will take place even if the nucleus is surgically removed, moreover the pattern of protein synthesis in such enucleated eggs is apparently normal.

These examples all demonstrate the infamous Comma Splice, in which a comma is mistakenly used to join what are really 2 separate sentences. Reading aloud, you should hear the material come to a complete stop before the words *however, therefore*, and *moreover.* Thus, you must replace the commas with either a semicolon or a period, as in these revisions of the first example:

> The brain of a toothed whale is larger than the human brain; however, the ratio of brain to body weight is greater in humans.

The brain of a toothed whale is larger than the human brain. However, the ratio of brain to body weight is greater in humans.

8. *concentration* and *density*. People often use *density* when they mean *concentration*, as in the following example:

   Larvae were more active at the highest of the 3 food densities.

   Although "density" *can* refer to the number of things per unit volume (e.g., cells per milliliter [ml], as in this example, or moles per liter), it can also mean mass per unit volume. To avoid ambiguity, it would be better to write about the 3 food "concentrations."

9. *varying* and *various*. *Varying* means "changing over time" or with changing circumstances, while *various* means "different." Consider the following example:

   We also examined feeding rates among animals maintained at varying temperatures.

   It is certainly possible that temperature was caused to change over time during the study, but in reality, the authors simply maintained animals at each of 4 constant temperatures. A revised version of the sentence might read:

   We also examined feeding rates among animals maintained at 4 temperatures over the range 15–29°C.

   Misuse of the word *varying* sometimes adds a bit of amusement to an otherwise dreary day:

   Five shells of varying sizes were then selected for each hermit crab.

   You can almost visualize each shell pulsating and undulating as it waits to be inspected by the hermit crabs. What the student meant to write, of course, was "various" or "different."

10. Proper use of commas in writing species names. Use commas to set off formal species names only when the formal names are preceded by specific common names. For example, you would use a comma before the species name here:

    We extracted genomic DNA from embryos of the common blue mussel, *Mytilus edulis*.

but not here:

> We extracted genomic DNA from embryos of the mussel *Mytilus edulis*.

11. Using scientific names as adjectives. According to the Council of Science Editors (formerly the Council of Biology Editors), it is acceptable to use scientific names as adjectives, but only in the following ways: You can talk about "streptococcal infections" or about "streptococcus infections" as long as the genus name (*Streptococcus*) is not capitalized and is not italicized or underlined (i.e., it's acceptable to use the generic name in the vernacular). You can use the formal genus name as an adjective only if you are referring to all species within the genus, as in "We included all known *Photinus* firefly species in our analysis." You can use an organism's complete scientific name (the binomial) as an adjective when you are referring only to members of that particular species, as in "*Photinus ignitus* fireflies were studied ..."
12. And don't forget: The data are ... (see p. 12).

A Grammatical Aside: Rules-That-Are-Not-Rules

> *[S]ome grammarians have invented rules they think everyone should observe.... But since grammarians have been accusing the best writers of violating these rules for the last 200 years, we have to conclude that for 200 years the best writers have been ignoring both the rules and grammarians.*
>
> JOSEPH WILLIAMS

Here are some of the rules-that-are-not-rules that Williams is referring to:

Never begin a sentence with And.

Never begin a sentence with But.

Never begin a sentence with Because.

Never write a paragraph with fewer than 3 sentences.

Never end a sentence with a preposition.

Never split an infinitive.

"To go boldly where no man has gone before?" I don't think so. You *can't* split an infinitive (e.g., to go, to see, to conquer) in Latin, or Spanish, or Italian, or French, because the infinitives in those languages are a single word. In English, they're 2 words, so they can be split. The renowned linguist

Steven Pinker puts it this way: "[F]orcing modern speakers of English to not split . . . an infinitive because it isn't done in Latin makes about as much sense as forcing modern residents of England to wear laurels and togas."*

Most of the rules-that-are-not-rules were taught to us in grade school for good reasons—for example, to keep us from making real grammatical mistakes, or to encourage us to develop our ideas more logically and fully. But they were training wheels: It's okay to take them off now! Certainly, it would irritate your readers to begin too many sentences with *but*, and you don't want to write so *cutely* that you give readers the impression you don't take your work seriously, but I just started a sentence with *but* one sentence back and you probably didn't even notice: Used every now and then, it's a very effective way of making an important point stand out.

A Strategy for Revising: Pass by Pass by Pass

As noted earlier, fixing one set of problems usually brings smaller problems to the surface. **Don't try to fix everything at once**. Instead, plan to make a series of passes through your work, fixing different sorts of problems with each pass. **The more time you allow for revision and the more revisions you complete, the more effective your writing will become**.

First, look carefully at content. Second drafts commonly arise from only a small portion of the first—perhaps a few sentences buried somewhere in the last third of the original. In such a case, you must abandon most of the first draft and begin afresh, but this time you are writing from a stronger base.

Now, read your paper again and look for organizational problems. Once you are happy with the order in which ideas are presented and have created convincing, logical connections between them, make another few passes through your paper and revise for completeness, conciseness, clarity, and flow. Use lots of strong verbs! Warning: As you improve the clarity and conciseness of the writing, you will sometimes find problems with the ideas themselves.

Finally, revise for spelling and grammatical mistakes. And don't forget to proofread carefully before you turn in your work.

BECOMING A GOOD REVIEWER

> *You don't always have to chop with the sword of truth. You can point with it, too.*
>
> ANNE LAMOTT

---

*From Pinker S. 1994. *The Language Instinct.* p. 386.

The best way to become an effective reviser of your own writing is to become a critical reader of other people's writing. Whenever you read a newspaper, magazine, or textbook, be on the lookout for ambiguity and wordiness, and think about how the sentence or paragraph might best be rewritten. You will gradually come to recognize the same problems, and the solutions to those problems, in your own writing.

Insist that fellow students give you drafts of their work to look over at least several days before the final piece is due. **Be concerned first with content.** Until you are convinced that the author has something to say, it makes little sense to be overly concerned with how he or she has said it, for the same reason it would make little sense to wash and wax a car that was headed for the auto salvage.

Take an especially careful look at the title and the first few paragraphs. Does the title indicate exactly what the paper or laboratory report is about? Do the title and first paragraph seem closely related? In the first 1 or 2 paragraphs, do the sentences flow logically, establishing a clear direction for what follows? Can you tell from the first or second paragraph exactly what this paper, proposal, or report is about, and why the issue is of interest? Or are you reading a series of apparently unrelated facts that seem to lead nowhere, or in many different directions? Does the first paragraph head in one direction, the second in another, and the third in yet another? If so, focus your comments on those issues.

If you are examining a laboratory report, study the Results section first. Does it conform to the requirements outlined in Chapter 9? Does the Materials and Methods section answer all procedural questions that were not addressed in the figure captions and table legends? Should some of those questions (e.g., experimental temperature) be addressed in the captions and legends, or directly on the graphs or in the tables? Does the Introduction section state a clear question and provide the background information needed to understand why that question is worth asking? Does it provide a compelling rationale for the work that was done? Does each sentence make sense, and does each lead in logical fashion to the next? Does each paragraph of the Introduction follow logically from the previous paragraph? Does the concluding paragraph address the issue raised in the first paragraph? Does the Discussion section interpret the data or simply apologize, and does the Discussion clearly address the specific issue raised in the Introduction? Are the author's conclusions supported by the data? Does the author suggest what issues should be addressed in future studies?

Only when you can answer yes to these questions should you worry about commenting on conciseness, completeness, grammar, and spelling. When examining a first draft, it may be most useful to write a few paragraphs of commentary to the author and not write directly on the paper at all. Don't feel compelled to rewrite the paper for the author;

your role is simply to point out strengths and perceived weaknesses and to offer the best advice you can about potential fixes. Here is an example of how this might be done; the student is making comments about the first draft of a research proposal written by a fellow student:

> Jim, I think you have a good idea for a project here, but it's not reflected in your introduction (or the title, but that can wait). The question you finally state in the middle of p. 4 caught me completely by surprise; at least until the bottom of p. 2 I thought you were interested in the effects of electromagnetic fields on human development, and by the end of p. 3, I wasn't sure *what* you were planning to study! On pp. 2–3 especially, I couldn't see how the indicated paragraphs (see my comments on your paper) related to the question you ended up asking. Or perhaps they *are* relevant, and you just haven't made the connections clear to me? The entire introduction seems to be in the "book report" format we discussed in class, rather than a piece of writing with a point to make (I'm having this trouble, too). The information you present is *interesting*, but a lot of it seems irrelevant. Try to make clearer connections between the paragraphs, perhaps by leaving some things out. As the Pechenik book says, "Be sure each paragraph sets the stage for the one that follows"—isn't that a great book? Here is a possible reorganization plan: Introduce the concept of electromagnetic fields in the first few sentences (what they are, what produces them); then mention potential damaging effects on physiology and development (at present it's not clear why the question is so important until one gets to p. 6!); then state your question and note why sea urchins are especially good animals to study. Will that work?
>
> Also in the Introduction, I would expand the paragraph on gene expression effects; discuss one or two of the key experiments in some detail, rather than just tell us the results. I think this is important, since *your* experiments are a follow-up on these.
>
> Table 1 seems redundant, since you present the same information in Figure 1. Delete Table 1.
>
> Your experimental design seems sound, although I'm not sure the experiments really address the exact question you pose in your introduction (see my comments on the draft; probably you just need to rephrase the question?). But I didn't see any mention of a control; without the control, how will you be sure that any effects you see are due to the electromagnetic field? Also, won't your treatment raise the water temperature? If so, you will need to control for that as well.
>
> Finally, you might want to ask Professor Cornell about this, but I think you should write for a more scientifically advanced audience.

> Your tone seems a bit too chatty and informal. And watch those prepositions—you use them almost as freely as I do! I enjoyed reading your paper and look forward to seeing the next draft!

Notice that this reviewer points out the strengths of the piece without overlooking the weaknesses and deals with the major problems first. **Be firm but kind in your criticism; your goal is to help your colleague, not to crush his or her ego.** Be especially careful to avoid sarcasm. Write a page of constructive criticism that you would feel comfortable receiving.

To help you give your classmates helpful criticism on drafts of their assignments, your instructor may provide you with, or ask you to develop, a peer-review sheet (for examples, see http://ase.tufts.edu/biology/faculty/pechenik/). You can also use such forms to self-criticize your own writing.

Receiving Criticism

> *I have probably learned more about the business of conducting research from referee comments than from any other single source.*
>
> PHIL CLAPHAM

Be pleased to receive suggestions for improving your work. A colleague who returns your paper with only a smile and a pat on the back does you no favor. It is good to receive *some* positive feedback, of course, but **what you are really hoping for is constructive criticism**. On the other hand, you need not accept every suggestion offered. Examine each one honestly and with distance, and decide for yourself if the reader is on target; reviews are advisory, giving you a chance to see how other people interpret what you have written. Sometimes a reader will misinterpret your writing, and you may therefore disagree with the specific criticisms and suggestions levelled at you. However, if something was unclear to one reader, it may be equally unclear to others. Try to figure out where the reader went astray, and modify your writing to prevent future readers from following the same path.

It is hard to read criticism of your writing without feeling defensive, but learning to value those comments puts you firmly on the path to becoming a more effective writer. After all, you *want* to communicate. If you are not communicating well, you need to know it, and you need to know why.

Fine-Tuning

Once the writing has a clear direction and solid logic, it is time to make 1 or 2 final passes to see that each sentence is doing its job in the clearest, most concise fashion. As a first step in developing your ability to fine-tune writing, read the following 26 sentences, and try to verbalize the ailment afflicting each one. Then revise the sentences that need help. Pencil your suggested changes directly onto the sentences, using the guide to proofreader's notation presented in Table 2 and the following example:

> Hermaphoditism is Commonly encountered among invertebrates. For example, the young East Coast oyster, Crassostrea virginica, matures as a male, later decomes a female and may chenge sex every few years there after. sequential hermaphrodites generally change sex only once, and usually change from male to female. In contraast to species than change sex as they age, many invertebrates are simultaneous hermaprhodites. Self-fertilization is rare among simultaneous hermaphrodites, it can occur, as in the tapeworms.

Table 2. Proofreader's symbols used in revising copy.

Problem	Symbol	Example
1. Word has been omitted	^ caret	study describes effect (with "the" inserted at caret)
2. Letter has been omitted	^ caret	that bok (with "o" inserted at caret)
3. Letters are transposed	∾	form the sea
4. Words are transposed	∾	was only exposed
5. Letter should be capitalized	≡ (three short underlines)	these data
6. Letter should be lowercase	/ (slash)	These Data
7. Word should be in italics	___ (underline once)	Homo sapiens

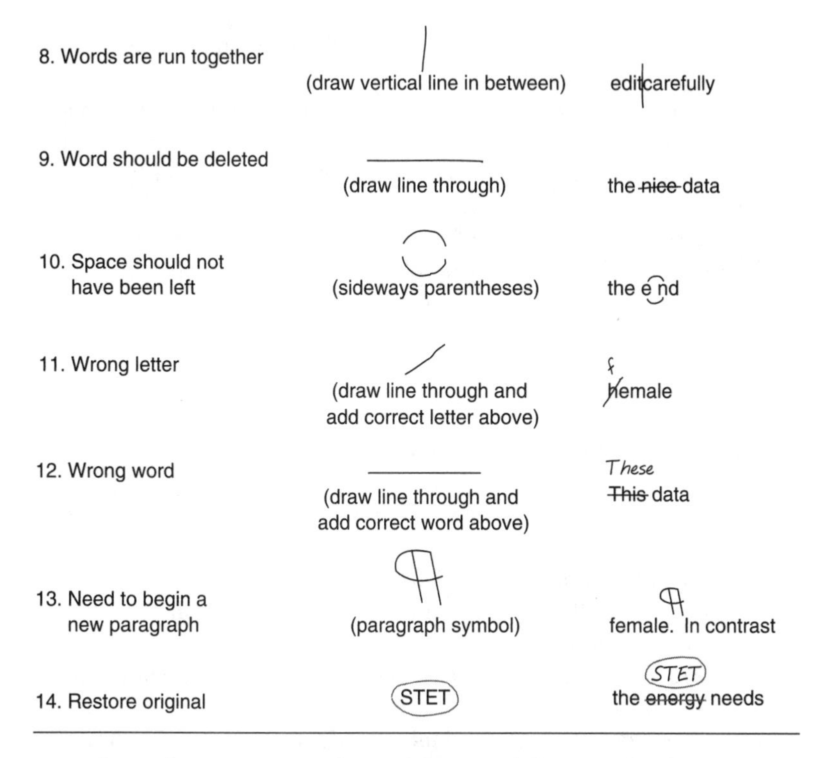

8. Words are run together	(draw vertical line in between)	edit|carefully
9. Word should be deleted	(draw line through)	the ~~nice~~ data
10. Space should not have been left	(sideways parentheses)	the e nd
11. Wrong letter	(draw line through and add correct letter above)	f ~~h~~emale
12. Wrong word	(draw line through and add correct word above)	These ~~This~~ data
13. Need to begin a new paragraph	(paragraph symbol)	female. ¶ In contrast
14. Restore original	STET	STET the ~~energy~~ needs

When editing someone else's work, use a different color pen or pencil to be sure that the author will see the suggested changes.

TECHNOLOGY TIP 5

Tracking changes made to documents

When co-authoring a paper, one person generally writes the first draft and other authors then suggest modifications. For group work, you might draft the Materials and Methods section, for example, while another student drafts the Results section. The Track Changes feature of Word allows you to email a draft of your section or paper (as an attachment) to another student and to then receive comments and suggested changes from that student directly on the document, and in ways that highlight the changes that were made.

To use this feature in Word, open the document, click Tools on the menu bar, and then select Track Changes. All subsequent additions to the text will appear in a different color, and deleted text will be indicated to the side.

You can Accept or Reject changes one by one or accept all changes in the document at once.

You can also add comments, suggestions, and questions using Insert and then Comment from the menu bar. Be sure to Accept All Changes and to delete all Comments before submitting the document to the instructor.

Don't send your draft to other students until you have revised it a few times yourself. The closer the draft is to final form, the more useful reviewers can be in giving suggestions.

Sentences in Need of Revision

1. To perform this experiment there had to be a low tide. We conducted the study at Blissful Beach on September 23, 2001, at 2:30 PM.
2. In *Chlamydomonas reinhardi*, a single-celled green alga, there are two matine types, 1 and 2. The 1 and 2 cells mate with each other when starved of nitrogen and form a zygote.
3. Protruding form this carapace is the head, bearing a large pair of second antennae.
4. The order in which we think of things to write down is rarely the order we use when we explain what we did to a reader.
5. The purpose of Professor Wilson's book is the examination of questions of evolutionary significance.
6. Swimming in fish has been carefully studied in only a few species.
7. One example of this capacity is observed in the phenomenon of encystment exhibited by many fresh water and parasitic species.
8. In a sense, then, the typical protozoan may be regarded as being a single-celled organism.
9. An estuary is a body of water nearly surrounded by land whose salinity is influenced by freshwater drainage.
10. The carbon-to-nitrogen ratio of the microbial films gives an indication of the film's nutritional quality. (Bhosle and Wagh, 1997).
11. In textbooks and many lectures, you are being presented with facts and interpretations.
12. The human genome contains about 25,000 genes, however there is enough DNA in the genome to form nearly 2×10^6 genes.
13. It should be noted that analyses were done to determine whether the caterpillars chose the different diets at random.

14. These experiments were conducted to test whether the condition of the biological films on the substratum surface triggered settlement of the larvae.
15. Various species of sea anemones live throughout the world.
16. This data clearly demonstrates that growth rates of the blue mussel (mytilus Edulis) vary with temperature.
17. Hibernating mammals mate early in the spring so that their offspring can reach adulthood before the beginning of the next winter.
18. This study pertains to the investigation of the effect of this pesticide on the orientation behavior of honey bees.
19. The results reported here have lead the author to the conclusion that thirsty flies will show a positive response to all solutions, regardless of sugar concentration (see figure 2).
20. Numbers are difficult for listeners to keep track of when they are floating around in the air.
21. Those seedlings possessing a quickly growing phenotype will be selected for, whereas. . . .
22. Under a dissecting microscope, a slide with a drop of the culture was examined at 50 ×.
23. Measurements of respiration by the salamanders typically took one-half hour each.
24. The results suggest that some local enhancement of pathogen specific antibody production at the infection site exists.
25. Usually it has been found that higher temperatures (30°C) have resulted in the production of females, while lower temperatures (22–27°C) have resulted in the production of males. (e.g., Bull, 1980; Mrosousky, 1982)
26. Octopuses have been successfully trained to distinguish between red and white balls of varying size.

There are several ways to improve each of these sentences. For reference, my revisions are shown in Appendix A, but you should make your own modifications before looking at mine. Be sure that you can identify the problem suffered by each original sentence, that you understand how that problem was solved by my revision, and that your revision also solves the problem (and does not introduce any new difficulties).

CHECKLIST

- ❑ Allow adequate time for revision (p. 81).
- ❑ Read your paper aloud slowly, and listen for problems as you read (pp. 81, 111).

- ❑ Don't worry about problems within individual sentences until your paper or report has a beginning, middle, and ending, with each idea leading logically into the next. Use idea maps to help organize your thoughts (pp. 81–85).
- ❑ Revise for content, clarity, and completeness:

  Make sure that each sentence says something of substance, and says what you intend it to say (pp. 85–89).

  Be cautious in drawing conclusions (p. 88).

  Keep an audience of interested peers in mind as you revise (p. 88).

  Use *it*, *they*, *their*, and other pronouns sparingly, and be sure they don't create ambiguity (pp. 92–95).

  Make each statement as specific as possible (pp. 94–96).
- ❑ Revise next for conciseness:

  Delete "It is interesting to note that … " and other running jumps. Just dive right into the issue being presented (pp. 87, 96–97).

  Eliminate unnecessary prepositions (pp. 98–99).

  Replace weak verbs with stronger ones (pp. 99–100).

  Try to have animals, enzymes, and molecules DOING something (p. 100–102).
- ❑ Revise next for flow (coherence): Improve the logical connections between sentences and paragraphs using appropriate transitional words, summary and repetition, and occasional semicolons (pp. 102–106).
- ❑ Incorporate definitions into sentences (p. 102).
- ❑ Eliminate teleology and anthropomorphism (pp. 10–11, 106).
- ❑ Proofread for spelling and grammatical errors (pp. 106–114).
- ❑ Turn in work that you are proud to have completed.
- ❑ Practice finding problems with other people's writing, including that in published papers (pp. 114–117).
- ❑ Give criticism to others that is substantive and honest but also constructive and not insulting (pp. 115–117).

Answer to short exercise in establishing coherence (p. 105): The correct sentence order is b-e-a-d-c.

II GUIDELINES FOR SPECIFIC TASKS

PRELUDE: Why are you writing papers and proposals and giving talks?

Every time you are asked to write a summary, a critique, an essay, a review paper, a lab report, or a research proposal, your instructor is committing himself or herself to many hours of reading and grading. There must be a good reason to require such assignments; most instructors are not masochists.

In fact, writing papers benefits you in several important ways. For one thing, you end up teaching yourself something relevant to the course you are taking. **The ability to self-teach is essential for success in graduate programs and academic careers, and it is in fact a skill worth cultivating for success in almost any profession.** In addition, you gain experience in reading the primary scientific literature, as discussed in Chapter 3. Textbooks and many lectures present you with facts and interpretations. By reading the papers on which those facts and interpretations are based, you come face to face with the sorts of data, and with the interpretations of data, on which the so-called facts of biology are based, and you gain insight into the true nature of scientific inquiry. **The data collected in an experiment are always real; interpretations, however, are subject to change.**

The writing assignments discussed in the rest of this book will help you move away from the unscientific, blind acceptance of stated facts and toward the scientific, critical evaluation of data and ideas. These assignments are also superb exercises in the logical organization, effective presentation, and discussion of information, all of which are skills that can only ease your career progress in the future. How fortunate you are that your instructor cares enough about your future to give such assignments!

There is one last reason that instructors often ask their students to do substantial writing in their courses. One can simply summarize a dozen papers in succession without understanding the contents of any of them:

Here is everything I found out about this topic. I call this the book report format, in which the writer merely presents facts uncritically: The authors did this; the authors did that; the authors found this; the authors suggested that. But by evaluating and synthesizing rather than simply paraphrasing and listing what others have already done, you can show your instructor that you really understand what you have read, that you have actually learned something and not simply memorized or mimicked the information presented to you. These skills are also essential for giving good talks, and, in turn, giving a good talk is excellent preparation for writing a strong paper. And the key to it all, as suggested in Chapter 3 (pp. 42–43), is first learning to summarize effectively.

7

WRITING SUMMARIES, CRITIQUES, ESSAYS, AND REVIEW PAPERS

Because you can't synthesize effectively until you can first summarize effectively, instructors often ask their students to write a short summary or a critique of one or more individual research articles early in the semester, as a way of preparing you to write a longer paper later in the semester. The skills I discuss in this chapter will also prepare you to prepare good research proposals, talks, posters, and the Introduction and Discussion sections of laboratory and other research reports. **Learning to summarize effectively opens many doors.**

What Lies Ahead? In This Chapter, You Will Learn

- How to write 4 major types of papers based on original research articles from the primary literature
- That a critique is basically a summary to which you add your own evaluations and suggestions
- That review papers are logical arguments, in which you set out to convince your readers of a particular point about an area of research
- Why you need at least several weeks to read and synthesize before you're ready to start writing your paper
- That a well-written first paragraph sets the stage for all that follows
- How to craft an informative title

WRITING SUMMARIES AND CRITIQUES

Assignments to write summaries and critiques ask you to read a paper from the original scientific literature (the "primary literature") and to then summarize or assess that paper, usually in fewer than 2 double-spaced, typewritten pages. *Brief* does not, in this case, mean *easy*. In fact, producing that 1- or 2-page summary or critique will probably require as much mental effort as that involved in preparing a full essay or review paper. Completing these assignments is no trivial matter. But preparing good summaries and critiques is an excellent way to push yourself toward true understanding of what you read—and of the nature of scientific inquiry.

To do well in these short assignments, you must fully understand what you have read, which usually means that you must read the paper many times, slowly and thoughtfully, as discussed in Chapter 3, and give yourself time to think.

Follow the same procedures whether you are asked to write a summary or a critique; indeed, a critique begins as a summary, to which you then add your own evaluation of the paper.

Writing the First Draft

You will know that you are ready to write your first draft of the assignment when you can distill the essence of the paper into a single, intoxicating summary sentence—or, at most, 2 summary sentences—as discussed in Chapter 3 (pp. 42–43; see also p. 127). These sentences should include *all* the key points, present an *accurate* summary of the study, be *fully comprehensible* to someone who has never read the original paper, and be in *your own words.* As a general rule, **do not begin to write your formal paper until you can write such an abbreviated summary**; this exercise will help you discriminate between the essential points of the paper and the extra, complementary details. Several examples of good summary sentences will be given later.

If you cannot write a satisfactory 1- or 2-sentence summary, reread the article; you'll get it eventually. Once your summary sentence is committed to paper, ask yourself these questions:

1. Why was the study undertaken? To answer this, draw especially from information given in the Introduction and Discussion sections of the paper.
2. What specific questions were addressed? Often you can see the essence of the question most clearly by studying a paper's figure and tables, as explained in Chapter 3 (pp. 37–41). Summarize each question in a single sentence.
3. How were these questions addressed? What specific approaches were taken to address each question on your list?
4. What assumptions did the authors make? Might any of them be wrong? Are they testable? How might they be tested?
5. What were the major findings of the study?

6. What was particularly interesting about the paper? The questions asked? Some aspect of the methodology? Some particular result or set of results? Some particular conclusion?
7. What questions remain unanswered by the study? These may be questions addressed by the study but not answered conclusively, or they may be new questions arising from the findings of the study under consideration.

Writing the Summary

When you can answer these questions without referring to the paper you have read, you are ready to write. Writing without looking at the original paper will help you avoid unintentional plagiarism (pp. 3–15) and will self-test your understanding of the paper. You can (and should) always go back to the original paper later to double-check and fill in specific factual details.

At the top of the page—below your name, the course designation, and the date—give the complete citation for the paper being discussed, beginning at the left-hand margin: names of all authors, year of publication, title of the paper, title of the journal in which the paper was published, and volume and page numbers of the article. On a new line, indent 5 spaces and begin your summary with a few sentences of background information. **Your introductory sentences must lead up to a statement of the specific questions the researchers set out to address**. Next, tell (1) what approaches were used to investigate each question and (2) what major results were obtained. **Present the results using the past tense**. Be sure to state, as succinctly as possible, exactly what was learned from the study.

To cover so much ground within the limits of one typewritten page is no small feat, but it can be done if you make certain that you fully understand what you have read. Consider the following example of a brief, successful summary. Before writing the summary, the student condensed the paper into these 2 sentences:

> The tolerance of a Norwegian beetle (*Phyllodecta laticollis*) to freezing temperatures varied seasonally, in association with changes in the blood concentration of glycerol, amino acids, and total dissolved solute. However, the concentration of nucleating agents in the blood did not vary seasonally.

Note that the two-sentence distillation contains considerable detail despite its brevity, implying impressive mastery of the paper's contents; it is complete, accurate, and self-sufficient. When you can write such sentences, pat yourself on the back and proceed; the hardest work is over.

SAMPLE STUDENT SUMMARY

Minnie Leggs
Bio 101
September 30, 2011

Van der Laak, S. 1982. Physiological adaptations to low temperature in freezing-tolerant *Phyllodecta laticollis* beetles. *Comp. Biochem. Physiol.* 73A: 613–620.

Adult beetles (*Phyllodecta laticollis*), found in Norway, are exposed to sub-zero (°C) temperatures in the field throughout the year. In general, organisms that tolerate freezing conditions either produce extracellular nucleating agents that trigger ice formation outside the cells rather than within them or they produce biological antifreezes, such as glycerol, that lower the freezing point of the blood and tissues to below that of the environment, thereby preventing ice formation. This study (Van der Laak, 1982) documents the tolerance of *P. laticollis* to below-freezing temperatures and determines how seasonal shifts in the temperature tolerance of these beetles are mediated.

Beetles were collected throughout the year and frozen to temperatures as low as −50°C; post-thaw survivorship was then determined. Determinations were also made of the concentrations of solutes in the blood (that is, blood osmotic concentration), total water content, amino acid and glycerol concentrations in the blood, presence of nucleating agents in the blood, and the temperature to which blood could be super-cooled before it would freeze.

The temperature tolerance of *P. laticollis* varied from about −9°C in summer to about −42°C in winter; this shift in freezing tolerance was paralleled by a dramatic winter increase in glycerol concentration and in total blood osmotic concentration. Amino acid concentration also increased in winter, but the contribution to blood osmolarity was small compared to that of glycerol. Nucleating agents were present in the blood year-round, ensuring that ice formation will occur extracellularly rather than intracellularly, even in summer.

For beetles collected in midwinter and early spring, blood glycerol concentrations could be artificially reduced by warming beetles to 23°C (room temperature) for about 24–150 h. When glycerol concentrations of winter and spring beetles were reduced to identical levels by warming, the spring beetles tolerated freezing better than the winter beetles; these differences in tolerance could not be explained by differences in amino acid concentrations. This result indicates that some other factors, as yet unknown, are also involved in determining the freezing tolerance of these beetles.

Analysis of Student Summary

The student has, within one typed page, successfully distilled a 7-page technical report to its scientific essence. Note that the student used the first 3 sentences to introduce the topic and then summarized the purpose of the research in one sentence. The next short paragraph summarizes the experimental approach taken, and the main findings of the study are then stated. No superfluous information is given; the product glistens with understanding. Rereading the student's 2-sentence encapsulation of the paper (p. 127), you can see that the student was, indeed, ready to write the report.

Note that the student correctly presented both the methods and the results in the past tense.

As a challenge to yourself, try writing a one-paragraph summary of the above example, cutting the length of the original summary by about 75%. **Summary is the ultimate test of understanding.**

Writing the Critique

A critique is much like a summary, except that you get to add your own assessment of the paper you have read. This does not mean you should set out to tear the paper to shreds; **a critical review is a thoughtful summary and analysis, not an exercise in character assassination**. Most biological studies have shortcomings, most of which become obvious only in hindsight. As I often tell my students, if we knew exactly what we were going to find, we would know exactly how to design the study! Yet every piece of research contributes some information, even when the original goals of the study were not attained. Emphasize the positive—**focus on what was learned from the study**. Although you should not dwell on the limitations of the study, you should point out these limitations toward the end of your critique. Were the conclusions reached by the authors out of line with the data presented? Do the authors generalize far beyond the populations or species studied? Which questions remain unanswered? How might these questions be addressed? How might the study be improved or expanded in the future? Keep this in mind as you write: You wish to demonstrate to your instructor (and to yourself) that you understand what you have read. Do not comment on whether you enjoyed the paper or found it to be well written; stick to the science unless told to do otherwise.

The Critique

Before writing the critique, the student produced this one-sentence summary of the paper.

> The egg capsules of the marine snails *Nucella lamellosa* and *N. lima* protected developing embryos against low-salinity stress, even though the solute concentration within the capsules fell to near that of the surrounding seawater within about 1 h.

Again, note that this one-sentence summary satisfies the criterion of self-sufficiency: It can be fully understood without reference to the paper it summarizes. The critique follows:

Saul Tee

Bio 101

April 2, 2010

Kînehcép, N.A. 1982. Ability of some gastropod egg capsules to protect against low-salinity stress. J. Exp. *Marine Biol. Ecol.* 63: 195–208.

The fertilized eggs of marine snails are often enclosed in complex, leathery egg capsules with 30 or more embryos being confined within each capsule. The embryos develop for one or more weeks before leaving the capsules. The egg capsules of intertidal species potentially expose the developing embryos to thermal stress, osmotic stress, and desiccation stress. This paper (Kînehcép, 1982) describes the ability of such egg capsules to protect developing embryos from low-salinity stress, such as might be experienced at low tide during a rainstorm.

Two snail species were studied: *Nucella lamellosa* and *N. lima*. Embryos were exposed, at 10–12°C, either to full-strength seawater (control conditions) or to 7–16% seawater solutions (seawater diluted with distilled water). The ability of egg capsules to protect the enclosed embryos from low-salinity stress was assessed by placing intact egg capsules into the test solutions for up to 9 h, returning the capsules to full-strength seawater, and comparing subsequent embryonic mortality with that shown by embryos removed from capsules and exposed to the low-salinity stress directly.

Encapsulated embryos exposed to the low salinities suffered less than 2% mortality, even after low-salinity exposures of 9 h duration. In contrast, embryos exposed directly to the same test conditions for as little as 5 h suffered 100% mortality. All embryos survived exposure to control conditions for the full 9 h, showing that removal from the capsules was not the stress killing the embryos in the other treatments. Sampling capsular fluid at various times after capsules were transferred to the diluted seawater, Kînehcép found that the concentration of solutes within capsules fell to near that of the surrounding water within about 1 h after transfer.

This study clearly demonstrates the protective value of the egg capsules of 2 snail species faced with low-salinity stress. However, Kînehcép was unable to explain how egg capsules of these 2 species protect the enclosed embryos since the capsules did not prevent the solute concentration of the capsular fluid from decreasing. Although Kînehcép plotted the rate at which the solute concentration fell within the capsules (his Fig. 1), he sampled only at 0, 60, and 90 min after the capsules were transferred to water of reduced salinity. I think he should also have sampled at frequent intervals during the first 60 min to discover how rapidly the solute concentration of the capsule fluid falls. As Kînehcép himself suggests, perhaps the embryos are less stressed if the concentration inside the capsule falls slowly. These experiments were all performed at a single temperature even though encapsulated embryos are likely to experience fluctuation in both temperature and salinity as the tide rises and falls during the day; the study should therefore be repeated using a range of temperatures likely to be experienced in the field. In addition, I suggest repeating these experiments using deep-water species whose egg capsules are never exposed to salinity fluctuations of the magnitude used in this study.

Analysis of Student Critique

As before, this student begins with just enough introductory information to make the point of the study clear and ends the first paragraph with a succinct statement of the researcher's goal. The methods and results of the study are then briefly reviewed, as in a summary. Whereas a summary would probably have ended at this point, the critique continues with thought-provoking assessments by the student. **Note that the student was careful to distinguish his thoughts from those of the paper's author** (see pp. 13–15, on plagiarism).

WRITING ESSAYS AND REVIEW PAPERS

Both short essays and longer review papers ask you to present critical evaluations of what you have read: The greater length of a review paper mostly reflects a more extensive treatment of a broader issue. In both cases, you synthesize information, explore relationships, analyze, compare, contrast, evaluate, and organize your own arguments clearly, logically, and persuasively, gradually leading up to an assessment of your own. A good review paper or short essay is a creative work; **you must interpret thoughtfully what you have read and write something that goes beyond what is presented in any single article or book consulted.** Your goal is to come up with new thoughts based on what others have already done and to persuade readers that your point of view makes sense.

In preparing an essay or review paper, you will go through the same processes that the writers of textbooks and review articles go through in presenting and discussing the primary research literature.

Getting Started

You must first decide on a general subject of interest. Often your instructor will suggest topics that have been successfully explored by former students. Use those suggestions as guides, but do not feel compelled to select one of those topics unless so instructed. Be sure to choose or develop a subject that interests you. It is difficult to write successfully about something that bores you.

To get started, you need only a general subject, not a specific topic. Stay flexible. As you research your selected subject, you will usually encounter an unmanageable number of pertinent references and must therefore narrow your focus. You cannot, for instance, write about the entire field of primate behavior, because the field has many facets, each associated with a large and growing literature. In such a case, you will find a smaller topic, such as the social significance of primate grooming behavior, to be more appropriate; as you continue your literature search, you may even find it necessary to restrict your attention to a few, or even only a single, primate species.

Alternatively, you may find that the topic originally selected is too narrow and that you cannot find enough information on which to base a substantial paper. You must then broaden your topic or switch topics entirely, so that you will end up with something to discuss. Don't be afraid to discard a topic on which you can find too little information.

Choose a topic you can understand fully. You can't possibly write clearly and convincingly on something beyond your grasp. Don't set out to impress your instructor with complexity; instead, dazzle your instructor with clarity and understanding. Simple topics often make the best ones for essays and papers.

Researching Your Topic

In writing this sort of paper, you are not setting out to mindlessly regurgitate what others have done and found, and you are not reading to memorize anything. Instead, **you want to evaluate and synthesize information from a number of sources to come up with a new way of looking at the field.** You are setting out to make a specific point (called a thesis statement) and to convince readers that your point is valid. To accomplish this, you will need to read your sources selectively and slowly, and think, think, think as you read. Begin by carefully reading the appropriate section of your textbook to get an overview of the general subject of which your topic is a part, and then consult at least one more specialized book before tackling the primary literature. Then locate and read research reports on your topic, following the advice presented in Chapter 3. Your goal is to select a small number of interrelated papers and to read these with considerable care and patience. You will win few points by accumulating a huge number of references that then receive only cursory attention.

Be sure you understand thoroughly what you have read. One of the best ways to self-assess your understanding is to summarize the material in your own words as you read along, paragraph by paragraph, section by section. When you finish reading each paper, try writing a 1-paragraph summary of what you have read, and then a 1- or 2-sentence summary, as illustrated on pp. 42, 127, 130. Then you will begin to see the relationships between the various papers as you continue your reading.

You should start researching your general topic (e.g., the use of snails as intermediate hosts in the life cycles of parasitic flatworms) many weeks before the first draft is due, both to give yourself time to read carefully and, especially, to give yourself time to digest the material that you read.

Developing a Thesis Statement

It takes a fair amount of mental effort to come up with a useful thesis statement. While taking notes, continually ask yourself:

- Why am I writing this down?
- What is especially interesting about this particular information?
- What puzzles me about what I have read?
- Can I see any relationship between this information and what I have already read, written, or learned?
- What assumptions do the authors make, and does each assumption seem reasonable and well supported?

Look also for apparent contradictions in the results of different studies, and in the interpretations of different authors. Look for patterns, and for exceptions to those patterns.

As you finish each paper, jot down some ideas for topics you would like to know more about. As you start developing opinions about what you have read, jot those thoughts down, too, and ask yourself what those opinions are based on. Writing down such thoughts will help self-provoke intellectual engagement with the material, an essential ingredient in the recipe for success.

Eventually, you will begin coming up with original ideas, interesting things you hadn't thought about before, and then you will be ready to draft a thesis statement that both you and your readers will find interesting.

A thesis statement needs support to be convincing. You might, for example, suggest that while many authors assume that all snail species serve as intermediate hosts for parasitic flatworms, there are good reasons to question this assumption. Or you might suggest that the remarkable diversity of reproductive patterns among marine animals is made possible by the chemical and physical properties of seawater. Or you might argue that frog tadpoles rely more on chemical cues than on visual cues to distinguish relatives from nonrelatives. Or you may not reach any definitive conclusion at all, and your argument might be just that we don't seem to understand as much about some particular topic as some people think we do, and that there are specific, important issues that need to be studied further.

Refine your draft thesis statement as you keep reading, and as you keep writing, until you have a statement that you find interesting, that is not self-evident, and that requires support. This statement will fuel the entire project. **You will present this statement near the beginning of your paper and then devote the rest of the paper to supporting it.**

There's something magical about thesis statements and where they come from. As you continue to read and think and write, they just mysteriously appear. If you start your assignment well in advance of when it is due and spend several hours a week reading and thinking about your topic—following the methods advocated in Chapter 3—I can almost guarantee that it will happen for you. If you wait to begin reading and thinking until a week or so before your report is due, I can almost guarantee that it will not.

Writing the Paper

Getting Underway: Taking and Organizing Your Notes

Some instructors add considerable work to their lives by monitoring your progress, such as asking to see your notes, lists of papers read, summaries or critiques of those papers (pp. 125–132), and partial drafts. This is splendid for you because it puts you on a schedule; otherwise, you must schedule yourself. **Start on the project as soon as possible,** and allocate at least a few hours a week to it, every week, until it's done.

Once you have at least a draft of a thesis statement, begin the formal writing process by reading all of your notes, preferably with hands off the keyboard and without pen or pencil in hand. Having read your notes to get an overview of what you have accomplished, reread them, this time with the intention of sorting your ideas into categories. Notes taken on index cards are particularly easy to sort, provided that you have not written many ideas on a single card; one idea per card is a good rule to follow. To arrange notes written on full-sized sheets of paper, some people suggest annotating the notes with pens of different colors or using a variety of symbols, with each color or symbol representing a particular aspect of the topic. Still other people simply use scissors to snip out sections of the notes and then group the resulting scraps of paper into piles of related ideas. And, of course, if you have entered notes directly into a computer, you can cut and paste the notes to group together those on related issues. Experiment until you find a system that works well for you.

At this point, you must **eliminate any notes that are irrelevant to the specific topic you have decided to write about.** No matter how interesting a fact or idea is, it has no place in your paper unless it helps you develop your argument. Some of the notes you took early on in your exploration of the literature are especially likely to be irrelevant to your essay, because these notes were taken before you had developed a firm focus. Put these irrelevant notes in a safe place for possible later use; don't let them coax their way into your paper.

You must next arrange your categorized notes so that your essay or term paper progresses toward some conclusion. Again, ask yourself whether a particular section of your notes seems especially interesting to you, and why it does, and look for connections among the various items as you sort. Idea mapping (Chapter 6, pp. 84–86) can be a great help in organizing your material.

The Crucial First Paragraph

The direction your paper will take must be clearly and specifically indicated in the opening paragraph, as in the following example written by student *A*:

Most bivalve mollusks are sedentary, living either in soft-substrate burrows (e.g., soft-shell clams, *Mya arenaria*) or attached to hard substrate (e.g., the blue mussel, *Mytilus edulis*) (Barnes, 1980). However, individuals of a few bivalve species live on the surface of substrates, unattached, and are capable of locomoting through the water. One such species is the scallop *Pecten maximus* (Thomas and Gruffydd, 1971). In this essay, I will argue that swimming is made possible in *P. maximus* by a combination of unique morphological and physiological adaptations, and I will then consider some of the evolutionary pressures that may have selected for these adaptations.

In this first paragraph, student *A* defines the topic, states the specific problem to be addressed, and tells us clearly why the problem is of interest: (1) the typical bivalve doesn't move and certainly doesn't swim,

(2) a few bivalves can swim, (3) so what is there about these exceptional species that enables them to do what other species can't? and (4) what forces may have selected for such swimming ability? Note that use of the pronoun *I* is now perfectly acceptable in scientific writing.

In contrast to the previous example, consider the following weaker (although not totally unacceptable) first paragraph written by student *B* on the same subject:

> Most bivalved mollusks either burrow into, or attach themselves to, a substrate. In a few species, however, the individuals lie on the substrate unattached and are able to swim by expelling water from their mantle cavities. One such lamellibranch is the scallop *Pecten maximus.* The feature that allows bivalves like *P. maximus* to swim is a special formation of the shell valves on their dorsal sides. This formation and its function will be described.

In this example, the second sentence weakens the opening paragraph considerably by prematurely referring to the mechanism of swimming. The main function of the sentence should be to emphasize that some species are not sedentary; the reader, not yet in a position to understand the mechanism of swimming, becomes a bit baffled. The next-to-last sentence of the paragraph ("The feature that allows . . .") also hinders the flow of the argument. Indeed, there doesn't seem to *be* any argument. This sentence summarizes the essay before it has even been launched, and again, the reader is not yet in a position to appreciate the information presented: What is this "special formation," and how does it have anything to do with swimming? The first paragraph of a paper should be an introduction, not a summary. It must set the stage for all that follows.

The last sentence of student *B*'s paragraph does clearly state the objective of the paper, but the reader must ask, "Toward what end?" The author has set the reader up for a book report, not for a critical evaluation or a persuasive argument. Reread the paragraph written by student *A*, and notice how the same information has been used so much more effectively, introducing a thoughtful essay rather than a tedious recitation of facts. Student *A*'s first paragraph was written with a clear sense of purpose; each sentence carries the reader forward to the final statement of intent, the argument on which the rest of the paper will be based. You might guess (correctly, as it turns out) from reading student *B*'s first paragraph that the rest of the paper was somewhat unfocused and rambling. In contrast, student *A*'s first paragraph clearly signals that what follows will be well focused and tightly organized. It might take 3 or 4 revisions, but be sure to get your papers off to an equally strong start.

The first paragraph of your paper must state clearly what you are setting out to accomplish and why. Every paragraph that follows the first paragraph should advance your argument, clearly and logically, toward the stated goal.

Supporting Your Argument

State your case, and build it carefully. Use your information and ideas to build an argument, to develop a point, to synthesize. **Sketching an idea map (Chapter 6, pp. 84–86) is an excellent way to organize your thoughts into a powerful and logical progression of ideas.** If you are writing an extensive term paper, you can (and should) use many of your idea map topics as headings or subheadings to help guide readers through the issues covered. *The Quarterly Review of Biology* and *Biological Reviews* provide particularly good examples of the effective use of headings and subheadings in structuring lengthy literature reviews.

Avoid the tendency to simply summarize papers one by one: The authors did this, then they did that, and then they suggested the following explanation. Instead, set out to compare, to contrast, to illustrate, to discuss. As described more fully in Chapter 5, you must back up all statements of fact or opinion with supporting documentation; this documentation may be an example drawn from the literature you have read or a reference (author and date of publication) to a paper or group of papers that support your statement, as in the following example:

> The ability of an organism to recognize "self" from "non-self" is found in both vertebrates and invertebrates (Kuby, 1997). Even the most primitive invertebrates show some form of this immune response. For example, Wilson (1907) found that cells disassociated from 2 different sponge species would regroup according to species; cells of one species never reaggregated with those of the second species.

Similarly, the opening sentences in the student examples on pp. 136 and 139 are supported by references. In contrast, the statements by student *B* on p. 137 have no such support, weakening considerably the writer's authority.

In referring to experiments, **don't simply state that a particular experiment supports some particular hypothesis**, or that a researcher reached a particular conclusion; *describe* the relevant parts of the experiment, and **explain how the results relate to the hypothesis in question**. For example, how potent are the following sentences?

Dudash and Carr (1998) presented evidence that deleterious recessive alleles were responsible for inbreeding depression in 2 closely related plants in the genus *Mimulus*. In contrast, Karkkainen (1999) showed that deleterious recessive alleles could not account for inbreeding depression in the self-incompatible herb *Arabis petraea*.

There is nothing in these sentences to convince readers of anything, not even that their author has read more than the title and abstract of the papers cited. In contrast, look at how much more convincing the following paragraph is:

Foreign organisms or particles that are too large to be ingested by a single leukocyte are often isolated by encapsulation, with the encapsulation response demonstrating clear species-specificity. For example, Cheng and Galloway (1970) inserted pieces of tissue taken from several gastropod species into an incision made in the body wall of the terrestrial gastropod *Helisoma duryi*. Tissue transplanted from other species was completely encapsulated within 48 hours of the transplant. Tissue obtained from individuals of the same species as the host was also encapsulated, but encapsulation was not completed for at least 192 hours.

In all of your writing, avoid quotations unless they are absolutely necessary; rely on your own words and your own understanding of what you have read.

The Closing Paragraph

At the end of your essay, summarize the problem addressed and the major points you have made, as in the following example:

Clearly, the basic molluscan plan for respiration that had been successfully adapted to terrestrial life in one group of gastropods, the terrestrial pulmonates, has been successfully readapted for life in water by the freshwater pulmonates. Having lost the typical molluscan gills during the evolutionary transition from salt water to land, the freshwater pulmonates have evolved new respiratory mechanisms involving either the storage of an air supply (using the mantle cavity) or a means of extracting oxygen while under water, using a gas bubble or direct cutaneous respiration. Further studies are required to fully understand how the gas bubble functions in pulmonate respiration.

Never introduce any new information in your summary paragraph.

Citing Sources

Cite only sources that you have actually read and would feel confident discussing with your instructor. Unless told otherwise, cite sources directly in the text by author and date of publication rather than by using footnotes. For example

> Kim (1976) demonstrated that magnetic fields established by direct current can alter the rates of enzyme-mediated reactions in cell-free systems. Similarly, magnetic fields established by alternating current can affect the activity of certain liver enzymes (Yashina, 1974) and mitochondrial enzymes (Kholodau, 1973).

More detailed information about citing sources is given in Chapter 5 (pp. 69–73).

At the end of your paper, include a Literature Cited section, listing all the publications referred to in your paper. Your instructor may specify a particular format for preparing this section of your paper. For specific information about preparing the Literature Cited section, see pp. 75–79.

Creating a Title

By the time you have finished writing, you should be ready to title your creation. **Give the essay or term paper a title that conveys significant information about the specific topic of your paper** (see also pp. 209–211):

> **No:** Factors controlling sex determination in turtles
>
> **Yes:** The roles of nest site selection and temperature in determining sex ratio in loggerhead sea turtles
>
> **No:** Biochemical changes during hibernation
>
> **Yes:** Adaptations to environmental stress: The biochemical basis for depressed metabolic rate in hibernating mammals
>
> **No:** The control of organ development in fish
>
> **Yes:** The novel gene "*exdpf*" regulates pancreas development in zebrafish

In your enthusiasm to make your title specific and informative, don't also make it unnecessarily wordy. How would you improve the following title?

> Does adaptation to copper result in a decrease in glutathione S-transferase activity over time in the marine mussel *Mytilus galloprovincialis*?

Following the Second Commandment of Concise Writing (Chapter 6, p. 99), let's strengthen the verb by replacing "result in a decrease" with simply "decrease." Also, doesn't "decrease" already imply the passage of time? By eliminating the redundancy, the title then becomes

> Does adaptation to copper decrease glutathione S-transferase activity in the marine mussel *Mytilus galloprovincialis*?

Revising

Once you have a working draft of your paper, you must revise it, clarifying your presentation, removing ambiguity, eliminating excess words, and improving the logic and flow of ideas. Revising is discussed in Chapter 6. If you are having difficulty organizing your ideas, I urge you to try idea mapping (pp. 84–86). You may also have to edit for grammar and spelling. Always leave time for at least 1 or 2 revisions of your work.

CHECKLIST FOR ESSAYS AND REVIEW PAPERS

- ❑ The title is informative.
- ❑ The opening paragraph indicates the specific direction that the paper will take, and it leads to a clear thesis statement that will drive the rest of the paper.
- ❑ All statements of fact and opinion are supported by references or examples.
- ❑ Research papers and other references are discussed in relation to one another, rather than in isolation.
- ❑ How particular results support particular hypotheses or lead to specific questions is indicated.
- ❑ No new information is presented in the final paragraph.

8

ANSWERING ESSAY QUESTIONS

BASIC PRINCIPLES

Answering essay questions on examinations is somewhat like writing summaries, except that you have only a short time to complete your work, usually from 15 to 50 minutes, and you have no choice in the subject of the essay. As with writing summaries, critiques, or review papers, you want your opening sentences to set the stage for all that follows, and, as in writing a review paper, you should discuss the components of your answer in relationship to one another. Here's an example of how this is done:

> A major difference in the physiological strategies employed by animals that live at freezing temperatures versus hot temperatures concerns the molecules that are synthesized in response to living at those temperatures. Cold-adapted organisms employ molecules that either lower the temperature at which water freezes within the organism or protect against the harmful effects of ice formation by causing water to freeze outside the tissues. In contrast, organisms adapted to hot temperatures prevent important cellular processes from failing by increasing the production of certain enzymes that stop, or reverse, protein denaturation.*

The author then goes on to expand these ideas. Note how he has very nicely set the stage for what is to come.

A winning answer to an essay question will also follow all the guidelines outlined in Chapter 1. In particular, make your statements specific, and support them with examples or diagrams. Your performance on essay questions can be strengthened by keeping in mind a few additional points:

1. **Read the question carefully before writing anything.** You must answer the question posed, not the question you would have preferred to see on the examination. In particular, note whether

*Based on an essay by Casey Diederich, 2011.

the question asks you to list, discuss, or compare. A list will not satisfy the requirements of a discussion or comparison. A request for a list tests whether you know all components of the answer; a request for a discussion additionally examines your understanding of the interrelationships among these components.

Consider this list of the characteristics of a Big Mac and a Whopper, based on a tax-deductible study conducted in Boston, Massachusetts, in January 2011:

BIG MAC	**WHOPPER**
2 beef patties	1 beef patty
patties 3.25" diameter	patty 3.75"–4.0" diameter
fried beef	broiled beef
3-part bun (3 slices)	2-part bun (top and bottom)
sesame seeds on top bun	sesame seeds on top bun
slice of pickle	slice of pickle
chopped onion	slices of onion
slice of cheese	2–3 slices of tomato
lettuce	ketchup
sauce	mayonnaise
$3.89	$3.89
packed in a card-board box	wrapped in waxed paper

Suppose you are asked to write an essay presenting the features of both items. Your essay might look like this:

> The Big Mac consists of 2 patties of fried ground beef, each patty approximately 3.25 inches in diameter, with lettuce, chopped onion, sliced pickle, a slice of cheese, some reddish sauce, and a 3-part bun, with the 2 patties separated from each other by one of the bun slices. The top slice of the bun is covered with sesame seeds. The Big Mac sells for $3.89 and is served in a cardboard box.
>
> The Whopper consists of one patty of broiled ground beef (approximately 4 inches in diameter), with mayonnaise, ketchup, several slices each of tomato, pickle, and onion, and a 2-part bun, with the upper half of the bun covered with sesame seeds. The Whopper sells for $3.89 and comes wrapped in waxed paper.

If you are asked to compare, or to compare and contrast, the 2 products, your essay must be written differently than the example above. All too often, when asked to "compare and contrast" *A* and *B*, students first write everything they know about *A*, then everything they know about *B*, and then conclude with something like, "And so you can see that *A* and *B* have many similarities and many differences." That is unacceptable. You are asking your instructor to make the comparisons! To make them yourself requires a thorough understanding of the information—an overview of the subject that goes beyond what is needed to simply list facts. The instructor's question is designed to see if you've gotten that understanding.

Here is an example, comparing the characteristics of the Big Mac and the Whopper:

> Both the Big Mac and the Whopper contain ground beef and are served on buns. The 2 hamburgers differ, however, with regard to the way the meat is cooked, the way the meat and bread are distributed within the hamburger, the nature of accompanying condiments, and how the sandwiches are served.
>
> The meat in the Big Mac is fried, and each sandwich contains 2 patties, each approximately 3.25 inches in diameter and separated from the other patty by a slice of bun. In contrast, the meat in the Whopper is broiled, and each sandwich contains a single, larger patty, approximately 3.75–4 inches in diameter. The top bun of both sandwiches is dotted with sesame seeds. Both the Big Mac and the Whopper contain lettuce, onion, and at least one slice of pickle. The Big Mac, however, contains chopped onion, whereas the onion in the Whopper is sliced. Moreover, the Big Mac has a slice of cheese. On the other hand, the Whopper comes with slices of only tomato. Both sandwiches contain a sauce: ketchup and mayonnaise in the Whopper and a premixed sauce in the Big Mac. The Big Mac and the Whopper are identically priced, at $3.89.

If you are asked for a comparison and respond with a list, you will probably lose points, not because your instructor is being picky but because you have failed to demonstrate your understanding of the relationship between the characteristics of the 2 products. **It is not the instructor's job to guess at what you understand**; it is your job to demonstrate what you know to the instructor. Note that the facts included are the same in the 2 essays. The difference lies in the way the facts are presented.

If asked for a list, give a list; this response requires less time than a discussion, giving you more time to complete the rest of the examination. When asked for a discussion, however, discuss: Present the facts, and support them with specific examples. When asked for a comparison, you will generally discuss similarities and differences, but the word *compare* can also mean that you should consider only similarities. Often an instructor will ask you to compare and contrast, avoiding any such ambiguity. If you have any doubts about what is required, ask your instructor during the examination.

2. **Present all relevant facts.** Although there are many ways to answer an essay question correctly, your instructor will undoubtedly have in mind a series of facts that he or she would like to see included. That is, the ideal answer to a particular question will contain a finite number of components; the way you deal with each of these components is up to you, but each of the components should be considered in your answer.

   Before you begin to write your essay, list all components of the ideal answer, drawing both from lecture material and from any readings you were assigned. For example, suppose you are asked the following question:

   > Discuss the influence of physical and biological factors on the distribution of plants in a forest.

   What components will the perfect answer to this question contain? Begin by making a list of all relevant factors as they occur to you—don't worry about the order in which you jot these factors down.

PHYSICAL	**BIOLOGICAL**
amount of rainfall	competition with other plants
annual temperature range	predation by herbivores
light intensity	
hours of light per day	
type of soil	
pesticide use	
nutrient availability	

This list is not your answer to the essay question; it is an organizing vehicle intended for your use alone. Feel free to abbreviate, especially if pressed for time ("nutr. avail.," "pred. by herbs"), but be certain you won't misunderstand your own notes while writing the essay.

In preparing to write your answer to the essay question, arrange the elements of your list in some logical order, perhaps from most to least important or so that related elements are considered together; this grouping and ordering are most quickly done by simply numbering the items in your list in the order that you decide to consider them. You have now outlined your answer; the most difficult part of the ordeal is finished.

Incorporate into your essay each of the ordered components in your list. Avoid spending all of your time discussing a few of these components to the exclusion of the others. If you discuss only 4 of the 8 relevant issues, your instructor will be forced to assume you don't realize that the other issues are also relevant to the question posed. Show your instructor that you know all the elements of a complete answer to the question.

3. **Don't waste time repeating the question** or making other such "running jumps"—just dive in with the answer (see example below). State your point of view, and then defend it.
4. **Stick to the facts.** An examination essay is not an exercise in creative writing and is not the place for you to express personal, unsubstantiated opinion. As with any other type of examination question, your instructor wishes to discover what you have learned and what you understand. Focus, therefore, on the facts, and as with all other forms of scientific writing, support all statements of fact or opinion with evidence or examples. You may wish to suggest a hypothesis as part of your essay; if so, be sure to include the evidence or logic upon which your hypothesis is based.
5. **Keep the question in mind as you write.** Don't include superfluous information. If what you write is irrelevant to the question posed, you probably won't get additional credit for your answer, and you will most likely annoy your instructor. If what you write is not only irrelevant but also wrong, you will probably lose points. By letting yourself wander off on tangents, you will usually gain nothing, possibly lose points, and probably lose your instructor's good will; certainly, you will waste time that might more profitably be applied elsewhere on the examination. Listing the components of your answer before you write your essay will help keep you on track.

APPLYING THE PRINCIPLES

To see how these principles are applied in a more realistic situation, consider this question: "Compare and contrast the locomotion of a mobile polychaete worm and a sea urchin (Figure 10)."

A good response might begin as follows:

> The locomotion of both polychaete worms and sea urchins involves many dozens of appendages that move in complex but highly coordinated patterns.

The student might add a few sentences here about the pattern of movement in both groups, and then continue as follows:

> However, the 2 types of animal differ in the skeletal systems they employ in moving the appendages, the manner in which the appendages form temporary attachments to the substrate, the degree of development of

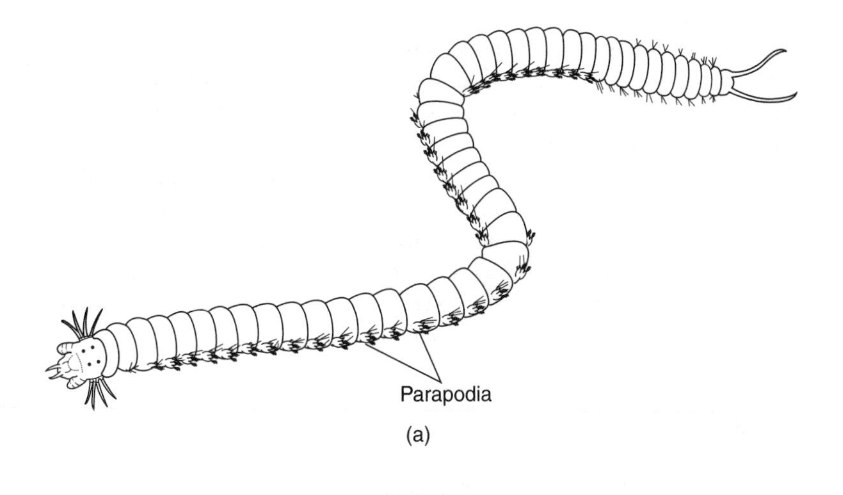

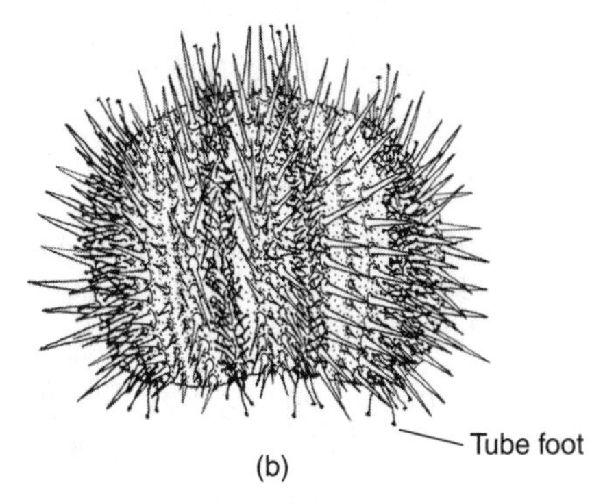

Figure 10. Sketches of the 2 animals discussed in the student's essay.
(a) a polychaete annelid and
(b) a sea urchin, showing locomotory appendages (parapodia and tube feet, respectively).

the nervous system directing and coordinating those movements, and the extent to which body wall musculature is also involved in locomotion. Whereas the movements of sea urchin appendages involve a fluid skeleton exclusively, the muscles operating polychaete parapodia act against each other through a rigid, internal skeleton, the acicula....

The answer would continue in this vein, dealing with the topics in the order they were listed in the second sentence. Note that the response begins with a specific answer, not with a "running jump"—the student does *not* write, "Polychaete worms and sea urchins have many similarities and differences in the way they locomote, which I will now discuss." Note also that the student has included no extraneous information, such as the names of the phyla the 2 animals are contained in, and that the student has listed *all* of the issues to be discussed early in the answer. You might want to leave a few lines of empty space in this part of your response so that you can add items that may occur to you as you keep writing or as you work on other parts of the examination. Even if this student runs out of time before completing a thorough discussion of each issue, he or she will earn substantial credit by having at least indicated all of the specific areas of overlap and contrast. Also, **note that every sentence makes some sort of comparison**, indicating an impressive command of the information. This student has been *processing* information, not just storing it. You can't help but do well on an essay exam, on any topic, if you follow this model.

SUMMARY

1. Your job is to show the instructor how much you understand; don't expect the instructor to guess.
2. Know the distinctions among listing, comparing, and discussing, and do exactly what is asked for.
3. When comparing or contrasting, make the comparisons explicit.
4. Make a quick side list of all components of the issue in question, mention each of those components in some logical order, and then fill in details as time permits.
5. Avoid running jumps: Don't rephrase the question; instead, dive right into the issue at hand.
6. Keep focused on the question you are expected to answer.

9

WRITING LABORATORY AND OTHER RESEARCH REPORTS

Inside every practical scientist the same pleasure in competing against nature lurks below the surface, the same enthusiasm for experiment, the same satisfaction in dreaming up new gambits and thrusts that can trick or tease the natural world into revealing its secrets.

JANET BROWNE

What Lies Ahead? In This Chapter, You Will Learn

- How to keep useful field and laboratory notes
- Why it is best to begin by writing the Materials and Methods and Results sections of your report
- How to determine what details to include in your Materials and Methods section
- That the centerpiece of your report is the Results section
- How to craft self-sufficient tables and figures
- How to choose the best way to present your data
- How to draw attention to the key features of your data
- How to write about the results of statistical analyses
- How to write an unapologetic, thoughtful Discussion section
- How to write a compelling Introduction—one that convinces readers about the importance of what you set out to test or understand
- The importance of writing an informative title, and how to do it
- How to write a brief and informative Abstract
- How to prepare and submit a paper for formal publication

WHY ARE YOU DOING THIS?

It is no accident that most biology courses include laboratory components in addition to lecture sessions. Doing biology involves making observations, asking questions, formulating hypotheses, devising experiments to test the hypotheses, presenting data, analyzing and interpreting data, and formulating new questions and hypotheses. **The so-called facts you learn from lectures and textbooks are primarily interpretations of data**. By participating in the acquisition and interpretation of data, you glimpse the true nature of the scientific process.

If you are contemplating a career in research, be assured that learning to write effective research reports now is an investment in your future. As a laboratory technician or research assistant, you will often be asked to analyze, summarize, and graph data so that the future path of the research can be decided. If you eventually pursue a research M.S. or Ph.D., you will find that a graduate thesis is essentially a very large lab report. Writing up your research for publication, either as a graduate student or as a researcher with a laboratory of your own, you will quickly find that you are again following exactly the procedures you used in preparing good laboratory reports in college biology courses; it can't hurt to learn the tools of your trade now.

You can also benefit from writing good research reports even if you do not expect to go on in the field of biology. Preparing such reports develops the ability to organize ideas logically, think clearly, and express yourself accurately and concisely. It is difficult to imagine a career in which the mastery of such skills is not a great asset.

THE PURPOSE OF LABORATORY AND FIELD NOTEBOOKS

The first step in preparing a good research report is keeping a detailed notebook. Research notebooks function to

1. Record the design and specific goals of your studies.
2. Record and organize your thoughts and questions about the work you are doing or planning to do.
3. Record your observations and numerical data.
4. Help you organize your activities in the laboratory or field so that you can work efficiently and accurately.

Keeping a detailed notebook will make the task of writing your report much easier, and you will end up producing a better report; after all, every product benefits from the use of quality materials. Moreover, the skills you learn in keeping the notebook could really pay off later in life. The Nobel Prize for the isolation of insulin would probably have gone to J. B. Collip instead of to Frederick Banting and C. H. Best had Dr. Collip learned to keep a more careful record of his work as a student. Collip was apparently the first to purify the hormone, but his notes were incomplete, and he was therefore unable to repeat the procedure successfully in subsequent studies.

Similarly, Charles Darwin neglected to make careful field notes about the birds he collected among the Galapagos Islands. Not realizing at the time that the birds collected on the different islands represented different species of finch, all of which were new to science, Darwin didn't even keep the specimens from different islands in separate packages. Fortunately, Captain FitzRoy and some of the crew members aboard the *Beagle* (including Darwin's much underacknowledged servant, Syms Covington) made their own bird collections during the voyage and *did* keep track of which birds came from which islands. Without their help, Darwin would never have been able to put the now-famous finch story together at all. You might not be so lucky: Take notes carefully and in detail.

Taking Notes

In keeping your research notebook, assume that you are doing something worthwhile, that you might well discover something remarkable (it does happen), and that you will suffer complete amnesia while you sleep that night. In other words, take the time to write down—in your own words—everything you are about to do, everything you actually do, and every reason why you do it. You can't always tell what will turn out to be important later on.

Record your data clearly, with each number identified by the appropriate units. Many of the details that seem too obvious to write down—for example, today's date (including the year), the name and location of your field site, the name of the species you are working with, or the units of measurement—are forgotten surprisingly quickly upon leaving the laboratory or field. You *can* write down too little, but it's difficult to write down too much.

Write so legibly and clearly that should you be run over by a truck on your way to class the next day, other students in the course would have no difficulty reconstructing your study and following your results. Similarly, if you use abbreviations in your notebook, be sure to indicate what each stands for. These procedures will greatly facilitate the writing of your research report. They are, in fact, crucial in any functioning research laboratory, because anyone in the lab must be able to pick up or interpret your work where you left off if you are unable to come in some day or if you leave that laboratory permanently.

Some of this writing can be done before the laboratory session. Whenever possible, read about the day's study ahead of time, being sure (through writing in your notebook) that you understand its goals, and plan exactly how you will record your data. You will get more out of the exercise and will undoubtedly finish your work in the lab or field more quickly if you arrive prepared. **It often helps to sketch a simple flowchart of your planned activities ahead of time**, as shown in Figure 11. (You can

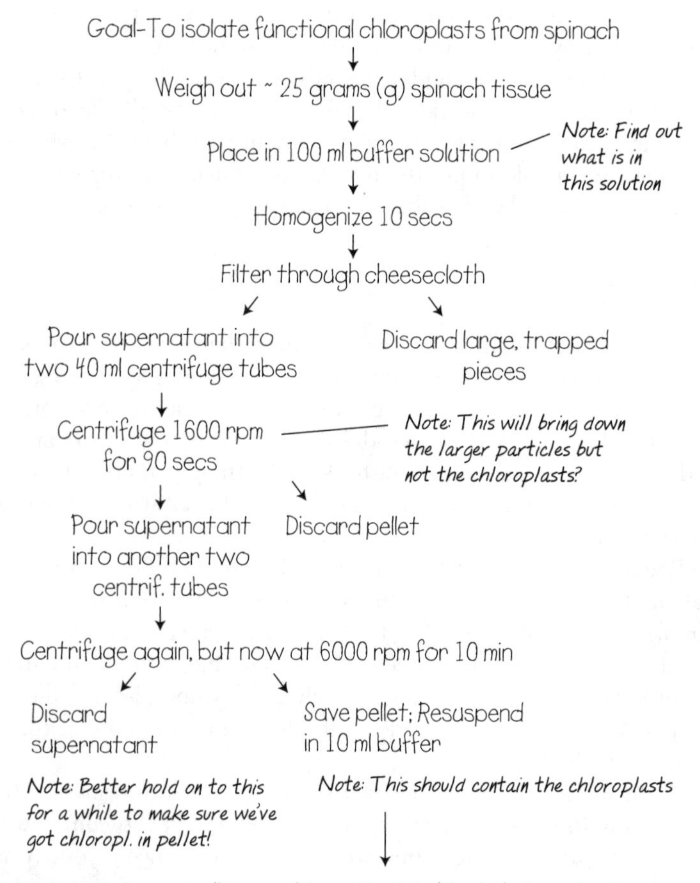

Figure 11. A page from a student's laboratory notebook. The flowchart is based on more detailed information provided by the instructor.

use this flowchart again later, in writing the Materials and Methods section of your report [pp. 158–164] or in preparing a poster or a talk [Chapter 11] about your work.)

Your research notebook should contain any thoughts, observations, or questions you have about what you are doing, along with the actual protocol and data. A sample page from a model notebook is shown in Figure 12. Note that the notebook entry begins by stating the date and

October 4, 2011

Goal—To determine rate at which the marine bivalve Mercenaria mercenaria (hardshell clam) moves water across its gills.

Approach—Use unicellular alga (phytoplankton) Dunaliella tertiolecta.

Determine initial cells · ml^{-1}, final cells · ml^{-1}. If know elapsed time and volume of seawater in container, can calculate cells eaten per hour per clam, and ml of sea H_2O cleared of cells/h/clam.

Weight of clam (incl. shell): 9.4g [Fisher/Ainsworth balance, Model MX-200]

Find initial concentration of algal cells:

1. 1 ml of culture + drop of Lugol's iodine to kill cells.
2. Load hemacytometer for cell counts—finally got it right on 5th try!
   Helps to tilt pipet at about 45° angle.

because of dilution w/0.05ml Lugol's sol'r

Data: (multiply by 1.05 and then by 10^4 → cells ml^{-1})

Note: ask Prof. Scully why 10^4.

counts per section of hemacytom slide

Sample 1	Sample 2
22	18
2) 13	21
18	20
4) 22	20
26	16
6) 21	12
20	18
8) 16	22
22	19
$\overline{X}$ = 20.0 (x 1.05) →	18.4 (x 1.05) →
→ 21.0 x 10^4 cells ml^{-1}	19.4 x 10^4 cells ml^{-1}

good agreement!

Put clam in 150 ml of this solution at 2:10 PM; H_2O = 21.4°C

Figure 12. A sample page from a student's laboratory notebook.

a specific goal. With such clear, well-organized, and well-thought-out notes, this student is well on his or her way to preparing a fine report.

A notebook that records field observations (a "field notebook") would look similar to that shown in Figure 13, which records part of a study

September 17, 2011. 1-3 PM. Sunny, 27°C air temp.; 17°C water.
Goal: Investigate size distribution of L. littorea shells at different distances from high tide line.
ME: Might expect shells to get larger as get closer to water, since these are marine snails and so should be able to feed more hours per day if spend more time under water. (What do they eat? Can they only eat in water?)
- - - -

Note: Many of the empty shells have large round holes made by a drilling predator—probably Nucella lapillus or Lunatia heros, which seem to be the only carnivorous gastropods living here.
→ Should calc. % of dead snails (empty shells) that have drill holes. Drill holes seem to be in exact same location on each shell. How is this possible?
- - - -

Note: Most of the rocks in this area are almost completely covered by very tiny (young) barnacles. Covering = so dense in some places can't even see the rock surface.
Question: Why no big, adult barnacles on these rocks?
Perhaps all die off during the summer for some reason.
There ~~are~~ many large barnacles on rocks much farther down the beach. Same species? If so, why only die here? Would be interesting to come back and monitor survival of young barnacles here and down the beach a few times during the summer and fall—perhaps once a month?
SAMPLING: 0.25 m^2 around each transect point, every 2 m from high tide (HT) mark.
DATA: No snails found from 0–3 meters (m) below high tide mark.
D = DRILLED

	Shell lengths (cm)	
Distance from high tide (m)	Live snails	
Empty shells		
4m	1.6, 1.4, 0.5	1.4
	1.7, 0.9, 2.1	
6m	1.7, 1.9, 1.1	1.8D
	2.0, 1.8, 1.8	1.4
		1.4

etc.

Figure 13. A sample page from a student's field notebook.

investigating the size distributions of the common marine intertidal snail *Littorina littorea* along a Massachusetts beach. Again, the quality of the entries suggests that this student is not simply trying to "get it over with," or simply writing down textbook facts and paraphrases of lecture notes; the student is clearly looking around and thinking—every instructor's dream.

Making Drawings

In both laboratory and fieldwork, it is often necessary to draw what you are seeing, particularly when working with living organisms. And perhaps the most important benefit is that **done properly, the act of drawing forces you to look more carefully at the object or organism before you**. As you draw, pay special attention to the relative sizes, shapes, and textures of different parts. Is part *A* connected to part *B*, or are they only adjacent to each other? Are the widths of parts *A* and *B* similar, or is one part wider than the other and, if so, how much wider? Is part *A* twice as wide as part *B* or 3 times as wide? The closer you look, the more you will see.

If possible, draw using a hard-lead pencil so that you can modify your drawings easily while you work. If you must use a pen, be sure the ink does not smear when wet. Try to figure out what things are as you draw, and label them as you go along. Be sure to indicate the approximate size of what you are drawing, and include a scale bar if appropriate. Most important, make your drawings *big* so that they can accommodate plenty of detail. Most beginning students make their drawings much too small; **think big**. Even if you are looking at something through a microscope, your drawing of that something should fill a 4-by-6-inch space. Remember, the goal is not to become a great artist; rather, the goal is to learn how to observe closely and how to record those observations accurately and in sufficient detail.

COMPONENTS OF THE RESEARCH REPORT

A research report is typically divided into 6 major sections:

1. *Abstract.* In the Abstract, you summarize the problem addressed, why the problem was addressed, your approach to the problem, and the major findings and conclusions of your study. This is probably the most difficult part of the report to write well; it summarizes the entire report, so save it for last.

2. *Introduction.* The Introduction, often only 1 or 2 paragraphs long, tells why the study was undertaken; a brief summary of relevant background facts leads to a statement of the specific problem being addressed. If appropriate, also describe the specific hypotheses that you set out to test, and the basis for those hypotheses.
3. *Materials and Methods.* This section is *your* reminder of what you did, and it also serves as a set of instructions for anyone wishing to repeat your study in the future.
4. *Results.* **This is the centerpiece of your report.** What were the major findings of the study? Present the data or summarize your observations using graphs and tables to reveal any trends you found. Point out major trends to the reader. If you make good use of your tables and graphs, the results can usually be presented in only 1 or 2 paragraphs of text; one picture is worth quite a few words. Avoid interpreting the data in this section.
5. *Discussion.* How do your results relate to the goals of the study, as stated in your Introduction, and how do they relate to the results that might have been expected from background information obtained in lectures, textbooks, or outside reading? Do your results support or argue against the hypotheses presented in your Introduction? What new hypotheses might now be formulated, and how might these hypotheses be tested? This section is typically the longest part of the report.
6. *Literature Cited ("References").* This section includes the full citations for any references (including textbooks, laboratory handouts, and Web sites) that you have cited in your report. Double-check your sources to be certain they are listed correctly; this list of citations will permit the interested reader to confirm the accuracy of any factual statements you make and, often, to understand the basis for your interpretations of the data. With only one exception (p. 72), cite only material you have actually read. The proper formats for citing literature and presenting citations are described in Chapter 5.

You may also be asked to include an Acknowledgments section, in which you formally thank particular people for their contributions to the project or to the report. And of course, your report will need an informative title.

Before writing your report, first study a few short papers in a relevant biological journal, such as *Biological Bulletin,*

Developmental Biology, or *Ecology*. Your instructor may provide you with a few especially good models. Reading these journal articles for content is unnecessary; you don't need to understand the topic of a paper to appreciate how the article is crafted. But do pay attention to the way the Introduction is constructed, the amount of detail included in the Materials and Methods section, and the material that is—and is not—included in the Results section.

While studying an article or two, note that figures and tables are always accompanied by explanatory captions (for figures) and legends (for tables), and that the axes of graphs and the columns and rows of tables are clearly labeled, with units of measure indicated. Note the location of figure captions and table legends. Study the captions and legends carefully, and imitate them (or improve them) in crafting your own.

WHERE TO START

Beginnings are hardest.... How can you write the beginning of something till you know what it's the beginning of?

PETER ELBOW

Strangely enough, you should not begin writing your report with the title, the Abstract, or the Introduction. It is far easier to write the Introduction—and the title—toward the end of the job, after you have fully digested what it is that you have done. And since the Abstract should be a tight summary of the entire report, this would be the worst section to begin with: How can you summarize something you haven't yet written? Save the Abstract (and the title) for later. **Start work with either the Materials and Methods section or the Results section.** Better still, you may profitably work on the two in tandem: Working on the Results section sometimes helps clarify what should be included in the Materials and Methods section, and working on the Methods sometimes clarifies the order in which results should be presented in the Results section.

Because the Materials and Methods section requires the least mental effort, completing it is a good way to overcome inertia. You may not know why you did the experiment or what you found out by doing it, but you can probably reconstruct what you did without much difficulty. Moreover, reminiscing carefully about what you did puts you in the right frame of mind to consider *why* you did it.

WHEN TO START

Start as soon as you can, preferably within 24 hours of finishing the study. In particular, start writing the Materials and Methods section while what you did is still fresh in your mind. Allow time for at least one major revision of each section (Chapter 6). When doing original research, try writing a detailed Materials and Methods section before you even conduct your study. This forces you to think about what you plan to do, how you plan to do it, and why you plan to do it, and can lead to improvements in experimental design before you actually start collecting data.

WRITING THE MATERIALS AND METHODS SECTION

Results are meaningful in science only if they can be obtained over and over, whenever the experiment is repeated. And because the results of any study depend to a large extent on the way the study was done, it is essential that you describe your methods so that your experiment can be repeated in all its essential details. Perhaps the best reason for writing a detailed Materials and Methods section is that it helps you review what you have done in an organized way and starts you thinking about why you have done it. Developing a good Materials and Methods section also puts you in the right frame of mind to do an equally good job on the other sections of the report.

The difficulty in writing this section of a research report (or journal manuscript) is in selecting the right level of detail, as discussed below. Students commonly give too little information; when informed of this defect, they may then give too much information. It's hard to hit it just right, but keeping your audience in mind (yourself and your fellow students) will help.

Determining the Correct Level of Detail

Many students begin with a one-sentence Materials and Methods section: "Methods were as described in the lab manual." Although this sentence meets the criterion of brevity, it is unacceptable as a stand-alone Materials and Methods section. For one thing, studies are rarely performed exactly as described in a laboratory manual or handout. Your instructions may call for the use of 15 animals, for example, but only

12 animals might be available for use on the day of your experiment. In addition, many details of a study will vary from year to year, from week to week, or from place to place and must therefore be omitted from your set of instructions.

Don't get carried away, however. Consider the following overly detailed description of a study involving the growth of radish seedlings:

> On January 5, I obtained 4 paper cups, 400 g of potting soil, and 12 radish seeds. I labeled the cups A, B, C, D and planted 3 seeds per cup, using a plastic spoon to cover each seed with about 1 cm of soil.

The author has used the first sentence simply to list the materials; whenever possible, it is far better to **mention each new material as you discuss what you did with it**. Furthermore, why do we need to know the weight of the soil obtained, or that the cups were labeled A–D rather than 1–4, or that a plastic spoon was used to add soil? Omitting the excess details and starting right in with what was done, we obtain the following:

> On January 5, I planted 3 radish seeds in each of 4 individually marked paper cups, covering the seeds with about 1 cm of potting soil.

Note that the essential details—individually marked cups, 3 seeds per cup, 1 cm of soil—not only survive in the edited version but now stand out clearly. The trick, then, is to determine which details are essential and which details are not. Happily, I've come up with a can't-fail method for getting this right.

Begin by listing all the factors that might have influenced your results. If, for example, you measured the feeding rates of caterpillars on several different diets, your list might look something like this:

Species of caterpillar used

Diets used

Amount of food provided per caterpillar

Time of year

Time of day

Air temperature in room

Manufacturer and model number of any specialized equipment used (e.g., balances, centrifuges, or spectrophotometers)

Size and age of caterpillars

Duration of the experiment

Container size or volume

Number of animals per container

Total number of individuals in the study

Precision of measurements made

Sometimes you will need to think carefully about whether a particular methodological detail might have influenced your results. Do you need to tell readers that you wore a red shirt the day you conducted the experiment? Not if you were studying the effect of different fertilizer concentrations on the growth rates of radish seedlings, but could the shirt have influenced your results if you were observing mating displays among sparrows? Possibly.

This list, which you do not turn in with your report, contains the bricks with which you will construct the Materials and Methods section. Each of the listed details must find its way into your report, although not necessarily in the order in which you jotted them down, because each gives information essential for later replication of the experiment. Some of this information may also help you explain why your results differed from those of others who have gone before you, a topic that will deserve some attention later, in the Discussion section of your report. Details that do not merit inclusion in the list are superfluous and should not appear in your Materials and Methods section.

In describing the procedures followed, **you must say what you did, but you should freely refer to your laboratory manual or handouts in describing how you did it**. For example, you might write

> The 3 diets were distributed to the caterpillars in random fashion, as described in the laboratory manual (Koegel, 2011).

The important point here is that the diets were distributed randomly; the outcome might be quite different if the largest caterpillars were to receive one diet and the smallest caterpillars another. The interested reader (including you, perhaps, at some later date) can refer to the stated source (Koegel, 2011) for detailed instruction in the method of randomization. You might want to append the relevant portion of your handout or manual at the end of your report as an appendix; this is a fine way to keep everything together for later use.

Be sure to note any departures from the given instructions. Suppose you were told to weigh the caterpillars individually but found that your balance was not sensitive enough to record the weight of a single animal. Your laboratory instructor, never at a loss for good ideas, probably suggested that you weigh the individuals in each container as a group. Your report might then include the following information:

> Determining the weight gained by each caterpillar over the 3-hour period of the experiment required that both initial and final weights be determined. The caterpillars were too small to be weighed individually. Therefore, similar-sized caterpillars were weighed in groups of 3 at a time, to the nearest 0.1 g. The average weight of each caterpillar in the group was then calculated.

Giving Rationales

Mention, for your own benefit as well as that of your readers, why particular steps were taken whenever you think it might not be obvious. Imagine yourself explaining things to a classmate who has not yet performed the study. We might, for example, profitably rewrite the sentence given on the previous page:

> To avoid prejudicing the results by distributing food according to size of caterpillar, the 3 diets were distributed to the caterpillars in random fashion as described by Koegel (2011).

A similar strategy can be used in writing about studies in molecular genetics:

> To test the sensitivity of *lig4* mutants to ionizing radiation, *lig4*^{169}/*FM7w* females were crossed to *lig4*^{169}/Y males.*

Describing Data Analysis

It is also usually appropriate to include any formulas used in analyzing your data. The following sentences, for example, would belong in a Materials and Methods section:

> The data were analyzed by a series of chi-square tests. The rate at which food was eaten was calculated by dividing the weight loss of the food by 3 hours, according to the following formula: Feeding rate = (Initial food weight − final food weight) ÷ 3 h.

Often this information is presented in a separate subsection at the end of the Materials and Methods section (see below).

---

*Modified from McVey *et al.*, 2004. *Genetics* 168: 2067–2076.

Note in this example that **"rates" always have units of "per time."** If it doesn't have units of "per time," it is not a rate (p. 110).

Use of Subheadings

Unless your Materials and Methods section is very short (e.g., a single paragraph), **use informative subheadings to help organize and present your material by topic**. The emphasis here is on the word *informative.* Below are some uninformative subheadings, relating to a study of shell choice by hermit crabs, followed by more substantive revisions on the same topic:

Uninformative: Field experiment

Informative: Occupancy of damaged and intact shells in the field

Uninformative: Shell choice

Informative: Effect of shell condition on shell choice in the laboratory

Two subsections commonly included at the end of the Materials and Methods section are "Data Analysis" and a description of your study system or organism. An example of a Data Analysis subsection is given on p. 163. Here is a sample subsection describing the study organism from a paper* reporting the role of environmental cues in stimulating larval metamorphosis:

Study Organism

Hydroides dianthus is a tube-dwelling serpulid polychaete found from New England through the West Indies, commonly on the underside of rocks (Hartman, 1969). Individuals are gonochoristic (i.e., they have separate sexes) and release gametes into seawater every 2–4 wk at 23°C (Zuraw and Leone, 1968). The larvae begin feeding 18–24 h after eggs are fertilized and become capable of metamorphosing after about 5 d at 23°C (Scheltema *et al.*, 1981; Bryan and Qian, 1997). The larvae undergo rapid and substantial morphological changes during metamorphosis that readily distinguish metamorphosed individuals from attached larvae (Scheltema *et al.*, 1981).

Alternatively, particularly if the organism or study site was chosen for specific attributes related to the nature of the research question, you could include this sort of information at the end of the Introduction (see p. 209).

---

*Modified from Toonen, R., Pawlik, J.R., 2001. *Mar. Ecol. Progr. Ser.* 224: 103–114.

Notice how the use of references in this example increases the author's credibility.

A Model Materials and Methods Section

The Materials and Methods section of your report should be brief but informative. The following example completely describes an experiment designed to test the influence of decreased salinity on the body weight of a marine worm. **Note that it is written in the past tense.**

Obtaining and Maintaining Worms

The polychaete worms used in this study were *Nereis virens*, freshly collected from Nahant, MA, and ranging in length between 10 and 12 cm. All treatments were performed at room temperature, approximately 21°C, on April 15, 2011. One hundred ml of full-strength seawater was added to each of six 200-ml glass jars; these jars served as controls, to monitor worm weight in the absence of any salinity change. Another 6 jars were filled with 100 ml of seawater diluted by 50% with distilled water.

Monitoring Water Gain and Loss

Twelve polychaetes were quickly blotted with paper towels to remove adhering water and were then weighed to the nearest 0.1 g using a Model MX-200 Fisher/Ainsworth balance. Each worm was then added to one of the jars of seawater. Blotted worm weights were later determined 30, 60, and 120 minutes after the initial weights were taken.

Determining Osmotic Concentration

The initial and final osmotic concentrations of all test solutions were determined using a Wescor VAPRO vapor pressure osmometer, following instructions provided in the handout (Podolsky, 2010).

Data Analysis

The rate of weight gain over time was examined by linear regression analysis, after log-transforming the independent variable (time). A series of Student's *t*-tests were used to assess the effect of salinity on rate of weight gain, by comparing mean weights of worms in the 2 treatment groups at 30, 60, and 120 minutes.

Note that all essential details have been included: temperature, species used, size of animals used, number of animals used per treatment, number of animals per container, volume of fluid in the containers, type and size of containers, time of year, equipment used, and how the data were analyzed. After reading this Materials and Methods section, you could repeat the study if you wanted to (or had to). Note, too, that **the writer has made clear why certain steps were taken**; 3 jars of full-strength seawater served as controls, for example, and worms were blotted dry to remove external water. The fact that worms were blotted dry before they were weighed was mentioned because that's a procedural detail that would obviously influence the results. On the other hand, the author does not describe how the balance and osmometer were operated since these techniques are standard, and the author does not tell us whether he or she used a clock or a stopwatch to monitor the passage of time because that choice could not have influenced the results obtained. The author has written a report that might be useful to someone in the future—and ends up with a top grade.

On to the Results!

WRITING THE RESULTS SECTION

The Results section is the most important part of any research report. Other parts of the report reflect the author's *interpretation* of the data. Interpretations necessarily reflect the author's biases, hopes, and opinions, and they are always subject to change, particularly as new information becomes available. In contrast, as long as a study was conducted carefully, and as long as the data were collected carefully, analyzed correctly, and presented accurately, the results are valid regardless of how interpretations change over time. Our understanding about how immune systems work has changed remarkably over the past decade, for example, and the current interpretation of older data differs considerably from the original interpretation. But the older data are still valid. The *results* of any study are real; the *interpretations* often change. That's why the Results section is indeed the centerpiece of your report.

In this section, you summarize your findings, using tables, graphs, and words. The Results section is

1. Not the place to discuss why the experiment was performed.
2. Not the place to discuss how the experiment was performed.
3. Not the place to discuss whether the results were expected, unexpected, disappointing, or interesting.

Simply present the results, drawing the reader's attention to the major observations and key trends in the data. Don't interpret them here. Most of the work in preparing this section of the report involves constructing a clear and well-organized presentation of data in the form of tables and figures.

What Is a "Figure"?

Note that in all of the examples given in this chapter, data summaries are formally referred to as either tables or figures. **"Figures" include graphs of all types; photographs of all types, whether of organisms or of electrophoretic gels; drawings; maps (showing the location of a study site, for example); and flowcharts. Anything and everything, in fact, that is not a "table" is a "figure." Most of your data will probably be presented in the form of tables and graphs.**

Summarizing Data Using Tables and Graphs

Before you even think about writing your Results section, you must work with your data. The observations you've made and the data you've collected most likely contain a story that is crying out for recognition. Contrary to popular opinion, the purpose of showing data in tables and graphs is not to add bulk to laboratory reports. Rather, **you wish to arrange the data in tables and graphs to reveal trends** and to make the most important points stand out, not only to your instructor but, more importantly, to yourself. The trick now is to organize the data so that (1) the underlying story is revealed and (2) the task of revealing the story to your reader is simplified.

There is no single right way to present data summaries; use any system that illustrates the trends clearly. First decide what relationships might be worth examining. Suppose we return to the experiment in which caterpillars fed for 3 hours on 3 diets. We determined both the initial weight of food provided and the weight of food remaining after the 3-hour period so that we could calculate the weight of food eaten per caterpillar per hour. In your report, you should provide a sample calculation so that if you make a mistake your instructor can see where you went wrong. We also know the initial and final weights of the caterpillars for each diet and the initial and final weights of dishes of food in the absence of caterpillars; these control dishes will tell us the amount of water lost from the food by evaporation.

What relationships in the data might be especially worth examining? The first step in answering this question is to **make a list of specific questions that might be worth asking**, as follows:

1. Did the caterpillars feed at different rates on the different diets? That is, did feeding rate vary with diet?
2. Did larger caterpillars eat faster than smaller caterpillars? That is, did feeding rate vary with caterpillar size?
3. How does the weight gained by a caterpillar relate to the weight lost by the food?
4. Did the weight of the control dishes change, and, if so, by how much?

The first question on this list was inspired by the null hypothesis established at the start of the study (in this case, H_0 = diet has no effect on feeding rate; see Chapter 4). But the other issues arose only after the data were collected and examined. **In scrutinizing your data, do not limit yourself to the questions and hypotheses posed at the start of your study.**

As in preparing the Materials and Methods section, this list of questions is for your own use and is not to be included in your report. Don't take any shortcuts here. Write these questions in complete sentences. Once you have this list of questions, it is easy to list the relationships that must be examined in your Results section:

1. Feeding rate as a function of diet
2. Feeding rate as a function of caterpillar size
3. Caterpillar weight gain versus food weight loss for each caterpillar
4. Food weight loss in the presence of caterpillars versus food weight loss in controls

Constructing a Summary Table

Now you must organize your data into a summary table in a way that will let you examine each of these relationships. Consider Table 3. This rough draft lists all the data obtained in the experiment. This summary table is for your own use; it is not usually something to be submitted as a formal part of your report. **Your goal here is to tidy up your data so that you can work with it more easily.**

For the first relationship in our list (feeding rate as a function of diet), a simple table will tell the entire story. For your report, you can simply present an abbreviated summary table, with explanatory caption, as in Table 4. Note that one of the caterpillars offered diet *A* ate no food

Table 3 Summary of raw data.

Diet	Initial Caterpillar Wt. (g)	Final Caterpillar Wt. (g)	Caterpillar Wt. Change (g)	Wt. of Food Lost (g) over 3 h	Feeding Rate (g food lost/h) of Caterpillar
A (Wheatgerm)	8.05	9.55	+1.55	3.65	$15.2 \cdot 10^{-2}$
A	4.80	5.80	+1.00	1.74	$7.2 \cdot 10^{-2}$
A	5.50	7.00	+1.50	3.33	$13.9 \cdot 10^{-2}$
A	5.50	4.70	~~−0.80~~	~~0.00~~	0
A	5.90	6.95	+1.05	1.35	$5.6 \cdot 10^{-2}$
Average	5.95	6.80	+1.28	2.52	$8.4 \cdot 10^{-2}$
B (sinigrin 10^{-5} M)	4.40	5.11	+0.71	2.19	$9.1 \cdot 10^{-2}$
B	5.20	5.60	+0.40	1.25	$5.2 \cdot 10^{-2}$
.	.	.	.	.	.
.	.	.	.	.	.
.	.	.	.	.	.
.	.	.	.	.	.
Control 1 (no caterpillars)	—	—	—	0.22	—
2	—	—	—	0.10	—
3	—	—	—	0.16	—

Table 4. The effect of sinigrin (allyl glucosinolate) added to a basic wheatgerm diet on food consumption of *Manduca sexta* caterpillars over 24 hours.

Diet	No. Caterpllars	(Mean g food eaten/ caterpillar/h)
Wheatgerm control	4*	$8.4 \cdot 10^{-2}$
Sinigrin (10^{-5} M)	5	$7.9 \cdot 10^{-2}$
Sinigrin (10^{-3} M)	5	$3.8 \cdot 10^{-2}$

**One individual died during the study without eating any food.*

and lost weight during the experiment. This individual died during the study, and the associated data were therefore omitted from Table 4. (The weight loss for this caterpillar probably reflects evaporation of body water.)

To Graph or Not to Graph

Finally, the time has come to reveal more subtle trends that may be lurking in the data. These trends may not be readily apparent from the summary table (Table 3); the trends may be made visible, however, both to you and your reader, through carefully designed and executed graphing. A word of caution: **Do not automatically assume that your data must be graphed.** If you can tell your story clearly using only a table—for example, if you are not interested in visualizing trends—a graph is superfluous. In other cases, you may be able to summarize some aspects of the data without using any graphs or tables at all. You might write, for example, "No animals ate at temperatures below 15°C" and then present data only for animals held at higher temperatures. **In any event, never present the same data in both a graph and a table**.

See Technology Tip 7 at the end of this chapter if you will be using Excel to plot graphs. In this chapter, a "thumbs up" symbol indicates a well-constructed, model graph, whereas a "thumbs down" symbol indicates a graph that suffers one or more major defects.

Graphs in biology generally take one of two basic forms: scatter plots (point graphs) or histograms and bar graphs. For the second relationship we wish to examine using the caterpillar data—feeding rate versus caterpillar size—a scatter graph, like the one shown in Figure 14, will be especially appropriate.

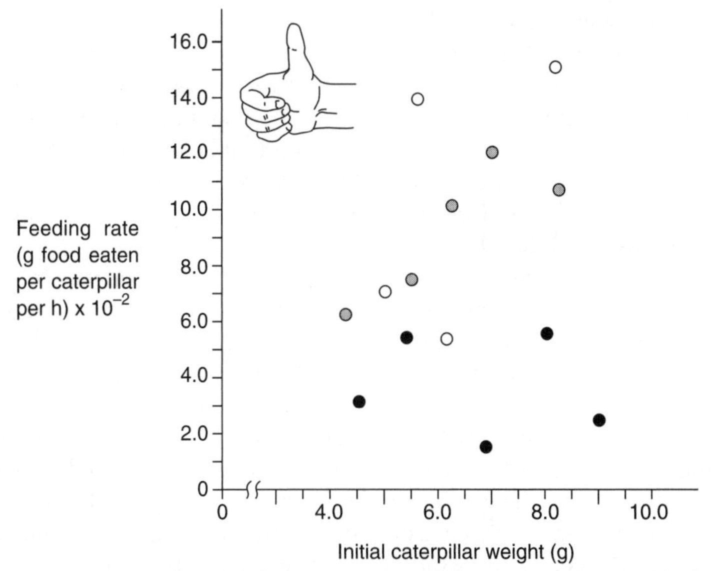

Figure 14. The relationship between initial caterpillar weight and rates of food consumption for *Manduca sexta* feeding for 3 hours on 1 of 3 diets at 24°C. Each point represents data from a different individual. Feeding rates varied from about 0.01 to 0.15 g eaten per caterpillar^{-1} h^{-1}. ○ = wheatgerm control diet; ● = 10^{-5} M sinigrin; • = 10^{-3} M sinigrin.

In examining Figure 14, note the following:

1. Each axis of the graph is clearly labeled and includes units of measurement.
2. Tick marks on both axes are at intervals frequent enough to allow readers to estimate the value of each data point.
3. The meaning of each symbol is clearly indicated.
4. **The symbols chosen facilitate interpretation of the graph:** Darker symbols represent increasing concentrations of sinigrin in the diet.
5. **The symbols are large and easy to tell apart.** Avoid using squares and circles of the same type (e.g., open square and open circle) on the same graph since these symbols can be difficult to tell apart.
6. A detailed explanatory caption (figure legend) is below the figure.

All of your graphs should exhibit these 6 characteristics. In Figure 14, it would be insufficient to simply label the *y*-axis "Feeding Rate." Feeding rates can be expressed as per minute, per hour, per day, or per year and as per animal, per group of animals, or per gram of body weight. Similarly, it is unacceptable to label the *x*-axis as "Weight," or even as "Caterpillar Weight." Don't make readers guess what you have done. From the figure caption, the axis labels, and the graph itself, the reader should be able to determine the question being asked, get a good idea of how the study was done, and be able to interpret the figure without referring to the text (see pp. 10 and 187). Never make the reader back up; **a good graph is self-contained**.

The third relationship (animal weight gain versus food weight loss) might well be left in table form since in this case the trend is readily discernible; caterpillars always gained less weight than that lost by the food (Table 3). The same trend could be revealed more dramatically (or, let us say, more graphically) with a scatter plot, as shown in Figure 15, but a graph is not essential here. Again, note the steps taken to avoid ambiguity: Axes are labeled, units of measurement are indicated, symbols are interpreted on the graph, and the figure is accompanied by an informative figure legend. Note also that the symbols used in the student's Figure 15 are consistent with their usage in Figure 14. **Always use the same system of symbols throughout a report** so as not to confuse your reader; if filled circles are used to represent data obtained on diet *A* in one graph, filled circles should be used to represent data obtained on diet *A* in all other graphs. And, as mentioned earlier, **try to use symbols that make sense whenever possible**. In Figures 14 and 15, increasing concentrations of sinigrin are represented by increasingly darker symbols.

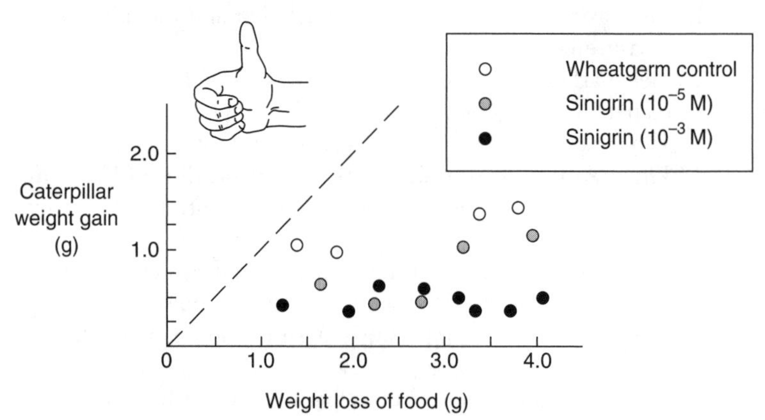

Figure 15. Caterpillar weight gain as a function of food consumption for *Manduca sexta* fed for 3 hours on 1 of 3 diets at 24°C. Points falling on the dotted line would indicate equality between weight gained and food eaten. Each point represents data from a different individual.

Similarly, if you were comparing metabolic rates at different times of day, you could help readers interpret your data, for example, by representing daytime measurements with open symbols and nighttime measurements with solid symbols. Note also that the order of symbols in the key corresponds to their placement in the figure, in which most of the open symbols lie above most of the filled symbols. Do whatever you can do to ease interpretation for the reader.

The fourth relationship in our list considers food weight loss in control dishes (no caterpillars). No graphs or tables are needed here; 2 sentences will do:

> Control containers exhibited less than a 3% weight loss (N = 3 containers) during the 24-h period. In contrast, food in containers with caterpillars lost at least 23% of initial weight.

If the weight loss had been substantial, perhaps 5% to 10% or more of initial weight, you might wish to adjust all the data in your tables accordingly before making other calculations:

> Control containers lost 7.6% of their initial weight (N = 3 containers) over the 24-h period. We therefore adjusted weight loss in other containers for this 7.6% evaporative loss before calculating feeding rates.

You would then provide a sample calculation, both for your instructor to examine and so that you will remember what you did if you consult your report again at a later date. A less desirable but nevertheless acceptable alternative would be to state the magnitude of the evaporative weight loss in your Results section and then bring this point up again when interpreting your results in the Discussion section. In this case, you would label appropriate portions of graphs and tables as "Apparent Feeding Rates" rather than "Feeding Rates." Again, although there are many wrong ways to present the data, there is no single right way; you must simply be complete, logical, consistent, and clear.

Note that in Figure 15, the student has defined the symbols directly in the graph rather than in the figure caption (compare with Fig. 14). When appropriate, you can also use empty space to report the results of statistical analyses, as in Figure 16, thus making the figure even more self-sufficient. Don't do this, however, if it will make the graph cluttered and difficult to read; instead, put the information in the caption and/or in the text of the Results section.

If the slope of the regression line had not been significantly different from zero, by the way, the student would have included neither the line nor the equation of the line in the figure (or elsewhere in the report).

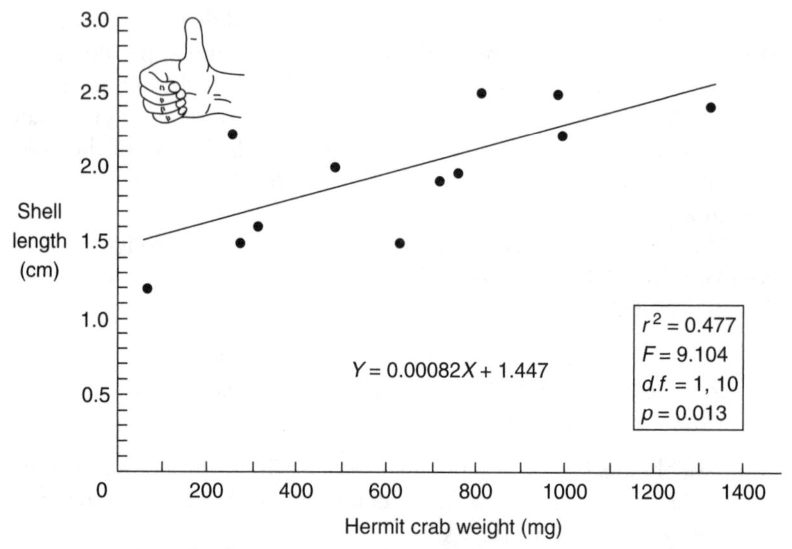

Figure 16. Relationship between wet weight (mg) of the hermit crab *Pagurus longicarpus* and the size of the periwinkle shells (Littorina littorea) occupied at Nahant, MA, on September 23, 2008 ($N = 12$). Each point represents a measurement on a single individual.

So far, we have looked only at examples of tables and point plots. If you were studying the differences in species composition of insect populations trapped in the light fixtures on 4 different floors of your biology building, a bar graph, as in Figure 17, might be more suitable. Note again that the axes are clearly labeled, including units of measurement, and that an explanatory legend accompanies the figure. Don't make the reader back up; make each figure self-sufficient. Note also that the graph tells an interesting story; given that *A* is the fruit fly *Drosophila melanogaster*, it is not difficult to guess where the genetics laboratory is located!

Use tables and graphs only if they make your data work for you; if a table or graph fails to help you summarize some trend in your results, it contributes nothing to your report and should be left out. Be selective. Don't include a drawing, graph, or table unless you plan to discuss it, and include only those illustrations that best help you tell your story.

Preparing Graphs

Graphs may be constructed with the aid of a computer (see Technology Tip 7), but unless your instructor suggests otherwise, don't feel that you *must* submit computer-generated graphs to earn a top grade. Most instructors would rather see a carefully thought-out, neatly executed

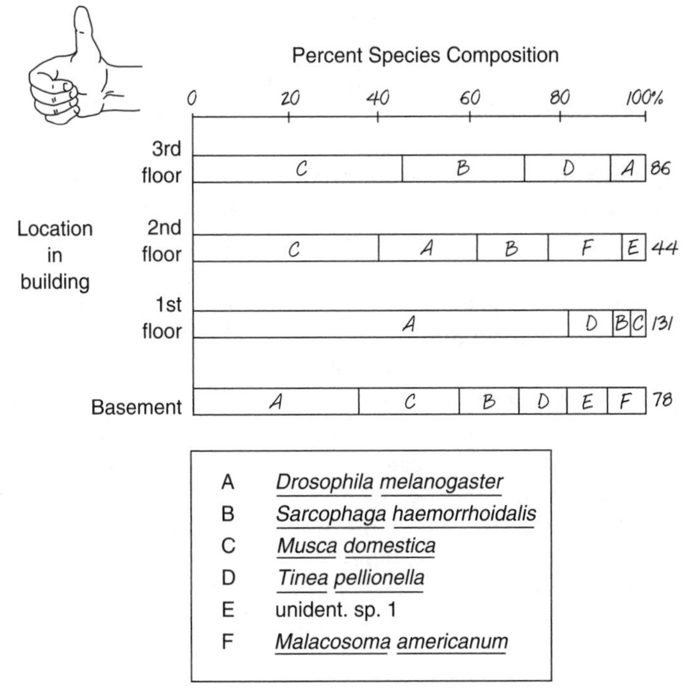

Figure 17. The distribution of insect species collected from light fixtures on 4 floors of the biology building on May 5, 2010. The number to the right of each bar gives the total number of insects collected on each floor.

graph done by hand than a poorly thought-out, neatly executed piece of computer graphics. To emphasize the point, I have retained many hand-drawn graphs in this book. I have seen some gorgeous pieces of complete garbage prepared using computers: I would rather see beginning students spend less time learning to use software and more time thinking about what they present, how they present it, and why they present it.

On the other hand, once you have mastered the key principles of graphing data, learning to use a good software package is certainly worthwhile, particularly since it allows you to quickly examine a variety of relationships in your data and determine which aspects of the data merit graphical presentation. But don't get carried away with all the bells and whistles; once the graphs are plotted, for example, it is sometimes faster to type or hand-print the axis labels or legends than to figure out how to have the computer execute these steps for you.

When preparing graphs by hand, always use graph paper, which can be purchased in your campus bookstore or downloaded online. The most useful sort of graph paper has heavier lines at uniform intervals—at every 4 or 5 divisions, for example, as shown in Figures 18 through 22. These heavy lines facilitate the plotting of data and reduce eye-strain, since every individual line need not be counted in locating data points.

By convention, the independent variable is plotted on the *x*-axis, and the dependent variable is plotted on the *y*-axis: *y* is a function of *x*. For example, if you examined feeding rates as a function of temperature (Fig. 18), you would plot temperature on the *x*-axis and feeding rate on the *y*-axis; feeding rate *depends* on temperature. On the other hand, temperature is not controlled by feeding rate; that is, temperature varies independently of feeding rate. Temperature is the independent variable and is plotted on the *x*-axis. Note that each point in Figure 18 represents data averaged from 5 individuals so that each point is an average value, or a "mean" value. This is clearly indicated in the *y*-axis label and in the figure caption.

It is a good practice to label the axes of graphs beginning with zero. To avoid generating graphs with lots of empty, wasted space, breaks can be put in along one or both axes, as in Figure 18 (see also Fig. 14 and

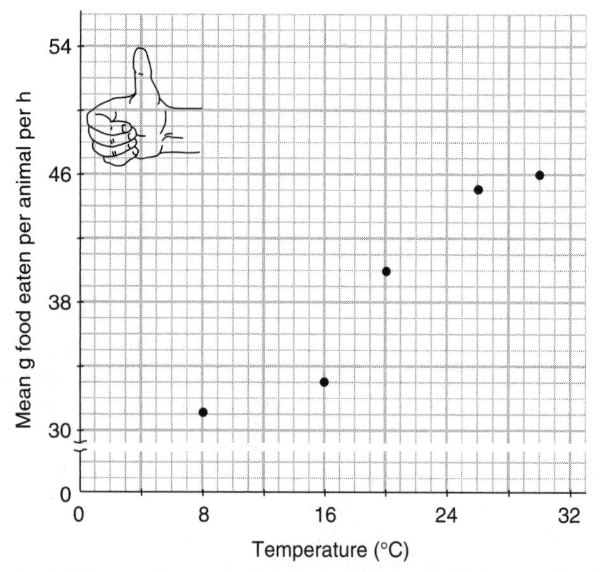

Figure 18. Feeding rate of *Manduca sexta* caterpillars on standard wheatgerm diet as a function of environmental temperature. Each point represents the mean feeding rate of 5 individuals measured over 24 hours.

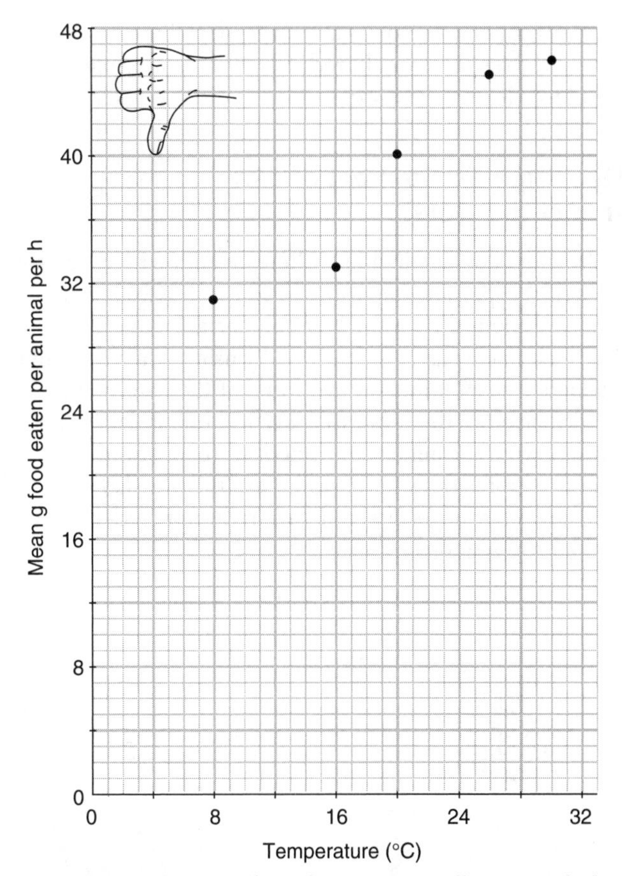

Figure 19. Feeding rate of *Manduca sexta* caterpillars on standard wheatgerm diet as a function of environmental temperature. Each point represents the mean feeding rate of 5 individuals measured over 24 hours.

Figs. 21–23). If a break had not been inserted in the *y*-axis of Figure 18, for example, the graph would have been less compact, with much wasted space, as in Figure 19.

(Not) Falsifying Data

You cannot move, add, or delete data points to improve or create trends in your data. Report the data you actually collected. **Falsifying data is, perhaps even more than plagiarism, an unforgivable offense**, and one that can get you into very serious trouble. Biologists build on the work of others, and that involves a lot of trust. Believe in the data that you collected, and present it unaltered. Sometimes unusual or unexpected results lead to interesting new questions or discoveries.

The Question: To Connect or Not to Connect the Dots?

After plotting data points, lines are often added to graphs to clarify trends in the data. It is especially important to add such lines if data from several treatments are plotted on a single graph, as in Figure 20. Note that this graph has been made easier to interpret by using different symbols for the data obtained at each temperature, and that the zero point on the *x*-axis has been displaced to the right, to prevent the first data point from lying on the *y*-axis, where it might be overlooked (compare with Figure 21, in which the first point does lie on the *y*-axis). Note also that each line is labeled. Using a key instead would have required readers to work a little harder in interpreting the graph.

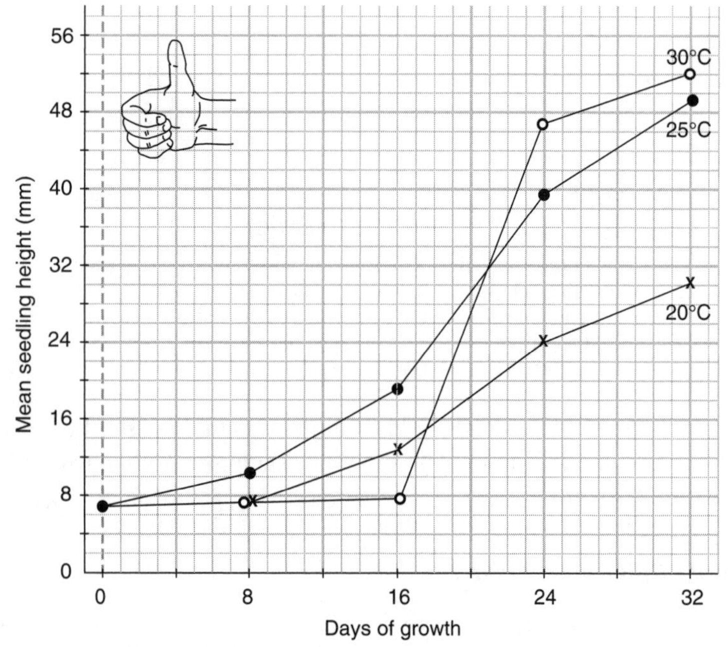

Figure 20. Rate of tomato seedling growth at 3 temperatures. Each point represents the mean height of 15 to 17 individuals.

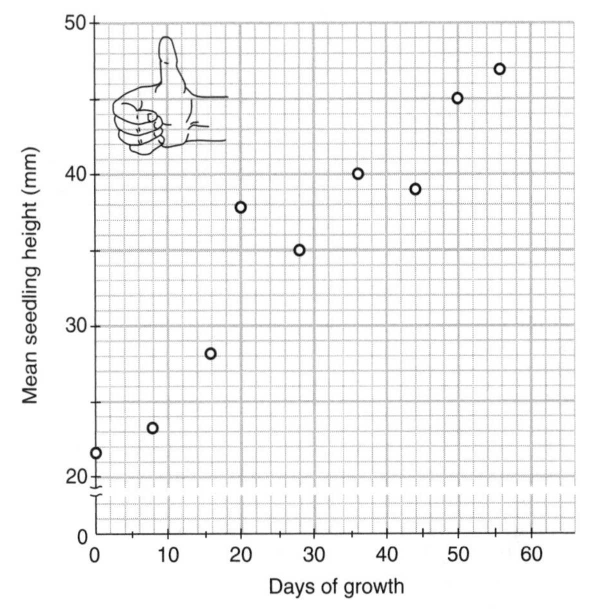

Figure 21. Rate of tomato seedling growth at 20°C. Fifteen to 20 seedlings, out of 205 seedlings in the population, were measured on each day of sampling.

In some cases, it makes more sense to draw smooth curves than to simply connect the dots. For example, suppose we have monitored the increase in height of tomato seedlings over some period of time in the laboratory. Every week, we randomly selected 15–20 seedlings from the laboratory population of several hundred to measure so that different seedlings were usually measured at each sampling period. After about 2 months, the data were plotted as in Figure 21.

Connecting the dots would not be the most sensible way to reveal trends in the data of Figure 21, because we know that the seedlings did not really shrink between days 20 and 28 or between days 36 and 44; simply connecting the points would suggest that shrinkage had occurred. The apparent decline in seedling height reflects the considerable variability in individual growth rates found within the same population, as well as the fact that we did not measure every seedling in the population on every sampling day. In this case, the trend in growth is best revealed by drawing a smooth curve, as shown in Figure 22.

When plotting average values (usually called arithmetic means) on a graph, **always include a visual summary of the amount of variation** present in the data by adding bars extending vertically from each point plotted (Fig. 22). You may, for example, choose to simply illustrate

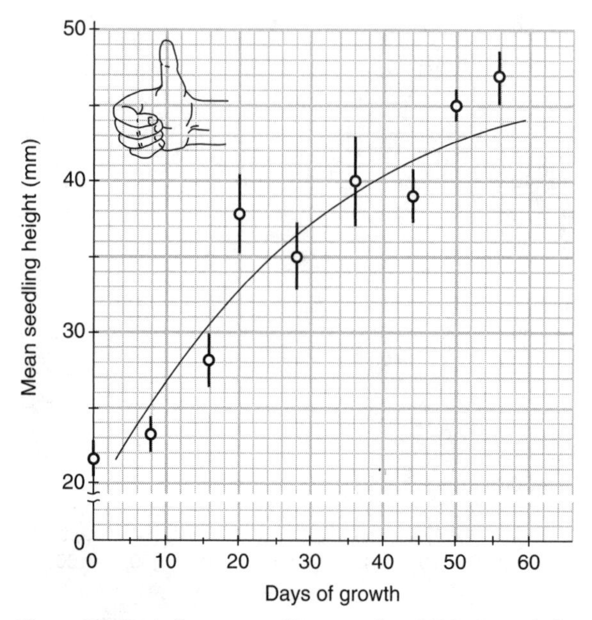

Figure 22. Rate of tomato seedling growth at 20°C. On each day of sampling, 15–20 seedlings were randomly sampled from a group of 205 seedlings and measured. Error bars represent one standard deviation about the mean.

the range of values obtained in a given sample. More commonly, you would plot "error bars" (typically the standard error or standard deviation about the mean), giving a visual impression of how much individual data points differed from the calculated mean values, as in Figure 22. The less overlap there is between error bars, the more likely it is that differences between mean values reflect real, biologically meaningful differences. Standard deviation and standard error calculations are reviewed in Chapter 4.

Plots of standard deviations or standard errors are always symmetrical about the mean value and so convey only partial information about the range of values obtained. If more of your individual values are above the mean than below the mean, the error bars will give a misleading impression about how the data are actually distributed. If your graph is fairly simple, you may be able to achieve the best of both worlds, indicating both the range and standard deviation (or standard error), as shown in Figure 23. In Figure 23, the bars extending vertically from the point at day 30 indicate that although the average seedling height was about 37 mm on that day, at least 1 seedling in the sample was as small as 25 mm, and at

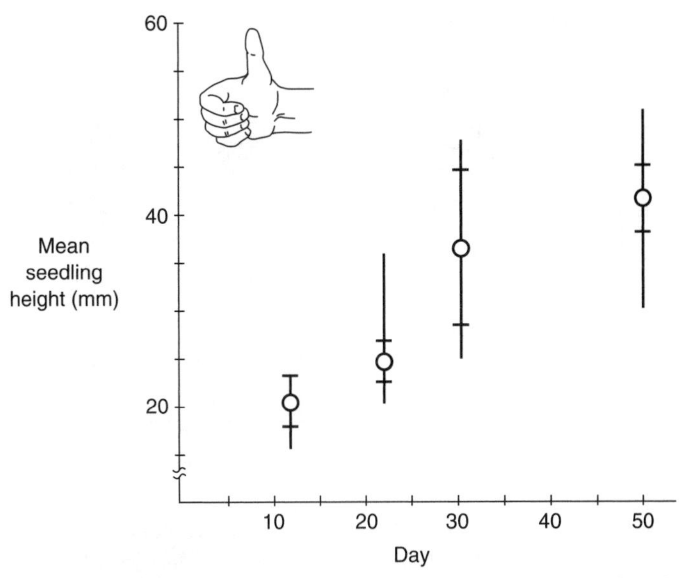

Figure 23. Mean height of tomato plant seedlings over a 50-day period. Each point represents the mean height of 18 seedlings raised at 20°C, with a photoperiod of 12L:12D. Vertical bars represent the range of heights; cross-bars represent 1 standard deviation about the mean.

least 1 seedling was as large as 48 mm. Seedlings measured on day 50 also differed in height by somewhat more than 20 mm. But we also see that the error bars are much larger for day 30 than for day 50, even though the range of lengths measured was similar for both samples. This tells us that most seedlings were close to the mean length on day 50, about 42 mm long, but that many of the seedlings measured on day 30 were substantially larger or smaller than the mean value measured on that day.

Whether you choose to plot standard deviations, standard errors, ranges, or some other indicator of variation, **be sure to indicate in your figure caption what you have plotted, along with the number of measurements associated with each mean**.

Making Bar Graphs and Histograms

When the variable along the *x*-axis (the independent variable) is numerical and continuous, points can be plotted and trends can be indicated by lines or curves, as we have seen in Figures 14–22. In Figure 18, for example, the *x*-axis shows temperature rising continuously from 0°C to 32°C, with each centimeter along the *x*-axis corresponding to a 4°C rise

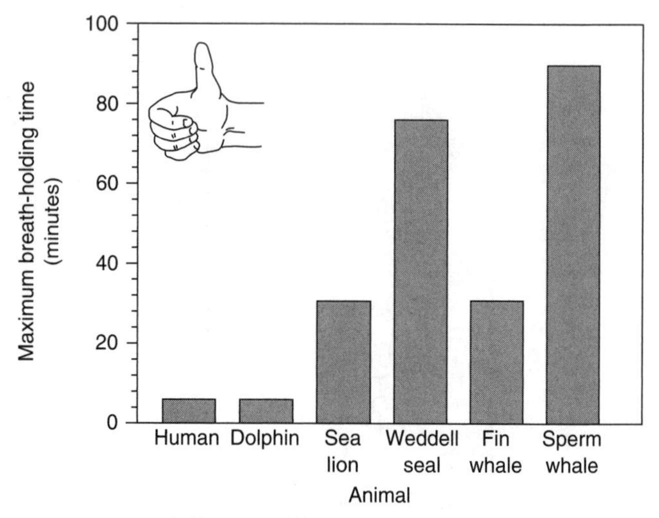

Figure 24. Breath-holding abilities of humans and selected marine mammals. *Data from Sumich, J.L.* 1999. An Introduction to the Biology of Marine Life, *seventh ed. WCB/McGraw-Hill, Publishers.*

in temperature. Similarly, the *x*-axes of Figures 21 and 22 reflect the march of time, from 0 to 60 days, with each centimeter along the *x*-axis reflecting 10 additional days.

When the independent variable is nonnumerical or discontinuous, or when the independent variable represents a range of measurements rather than a single measurement, the data are represented by bars, as shown in Figures 24 and 25. The *x*-axis of Figure 24 (a bar graph) is labeled with the names of different mammals. In contrast to the *x*-axes of Figures 14–23, the *x*-axis of Figure 24 does not represent a continuum: No particular quantity continually increases or decreases as one moves along the *x*-axis, and a line connecting the data for sea lion and Weddell seal would be meaningless. In Figure 25 (a histogram), the data for shell length are numerical but are grouped together (e.g., all shells 25.0–29.9 mm in length are treated as a single data point). Note also that the magnitude of the size categories represented by the different bars varies: The leftmost bar represents the percentage of shells found within a range of about 0.1–21 mm in length, whereas each of the next several bars to the right represents the percentage of shells found within a range of only about 5 mm in length. The size range of shells represented by the bar at the extreme right side of the graph is unknown. We know that all shells found in this category exceeded 45 mm in length, but the graph does not indicate the size of the largest shell.

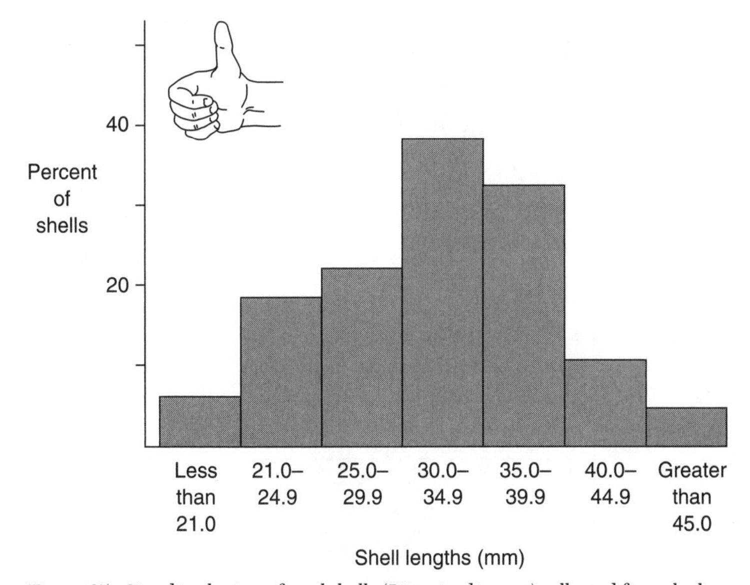

Figure 25. Size distribution of snail shells (*Littorina littorea*) collected from the low intertidal zone at Blissful Beach, MA, on August 15, 2010. Only living animals were measured. A total of 197 snails were included in the survey.

Use a single color or fill pattern for all bars unless there is a logical reason not to do so. In Figures 24 and 25, for example, all bars should be the same color or fill pattern.

Learning to Love Logarithms

Logarithmic scales make excellent sense when the data presented cover 2 or more powers of 10. Consider the following example: We wish to explore the relationship between the size of hermit crabs and the mass of the snail shells that they choose to live in. The relationship for the tropical hermit crab species *Clibanarius longitarsus* is presented using a standard linear scale in Figure 26*a*. The crabs measured ranged in mass from 0.3 g to nearly 40 g (2 powers of 10). Notice that data points for many quite small crabs are crowded together at the lower left side of the graph so it is difficult to tell the actual values for either crab mass or shell mass, or even to tell how many data points are represented. In addition, the data point for the largest (heaviest) crab is easily overlooked, since it is the only data point for any crab heavier than about 10 g (grams). Plotting logarithms (base 10) of the same data on the *x*-axis (Figure 26*b*) creates a more uniform distribution of data points. Individual data points are now much

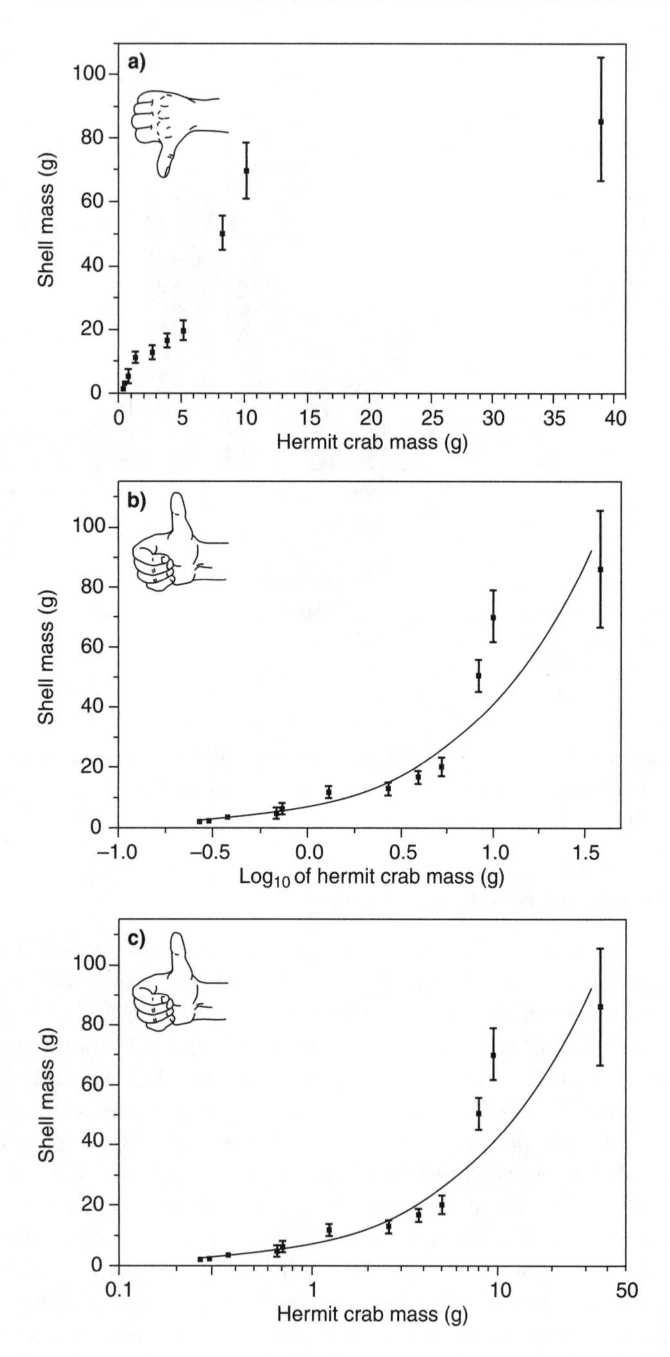
a)
Shell mass (g)
0
20
40
60
80
100
0
5
10
15
20
25
30
35
40
Hermit crab mass (g)
b)
Shell mass (g)
0
20
40
60
80
100
–1.0
–0.5
0.0
0.5
1.0
1.5
Log10 of hermit crab mass (g)
c)
Shell mass (g)
0
20
40
60
80
100
0.1
1
10
50
Hermit crab mass (g)

Table 5. Understanding logarithms. (a) In log base 10, each successive whole number is 10 times larger than the preceding whole number. Log base 10 scales are especially useful in presenting data that cover many powers of 10. (b) In log base 2, each successive whole number is twice as large as the preceding whole number. Log base 2 scales are especially useful for visualizing doublings of quantities. Note that in both systems, the log of a number is simply the exponent to which the base (10 in "a," 2 in "b") must be raised to obtain that number.

(a) Number	Log base 10	(b) Number	Log base 2
$0.01 = 1/100 = 10^{-2}$ →	(−2)	$1/4 = 2^{-2}$ →	(−2)
$0.1 = 1/10 = 10^{-1}$ →	(−1)	$1/2 = 2^{-1}$ →	(−1)
$1 = 10^{0}$ →	(0)	$1 = 2^{0}$ →	(0)
$10 = 10^{1}$ →	(1)	$2 = 2^{1}$ →	(1)
$100 = 10^{2}$	$4 = 2^{2}$	2	
$1{,}000 = 10^{3}$	3	$8 = 2^{3}$	3
$10{,}000 = 10^{4}$	4	$16 = 2^{4}$	4
$100{,}000 = 10^{5}$	5	$32 = 2^{5}$	5

easier to see. However, this representation is still somewhat difficult to interpret, because it requires you either to know how to decode logarithms or to refer to Table 5. Alternatively, we can replot the data of Figure 26*a* using a log base 10 scale on the *x*-axis, with each major tick now representing 10 times the value of the preceding numbered tick. The value 0 on the *x*-axis of Figure 26*b* corresponds to the value 1 on the *x*-axis of Figure 26*c*, while the value 1.0 on the *x*-axis of Figure 26*b* corresponds to the value 10 on the *x*-axis of Figure 26*c* (see Table 5).

←

Figure 26. The relationship between the mass of hermit crabs (*Clibanarius longitarsus*) from Mozambique and the mass of the shells they choose to occupy. (a) The data are plotted on a normal, linear scale. Note the large number of data points crowded together in the lower left corner of the graph. (b) The same data replotted after taking the log (base 10) of hermit crab mass. A value of 1 represents 10 g (grams) (see Table 5). (c) The same data plotted using a logarithmic scale (base 10) for the x-axis. Hermit crab mass increases 10-fold from one numbered tick mark to the next. In both (b) and (c), note the more even distribution of points. *Based on data of Barnes, D.K.A., 1999. Ecology of tropical hermit crabs at Quirimba Island, Mozambique: shell characteristics and utilization*. Mar. Ecol. Progr. Ser. 183: 241–251.

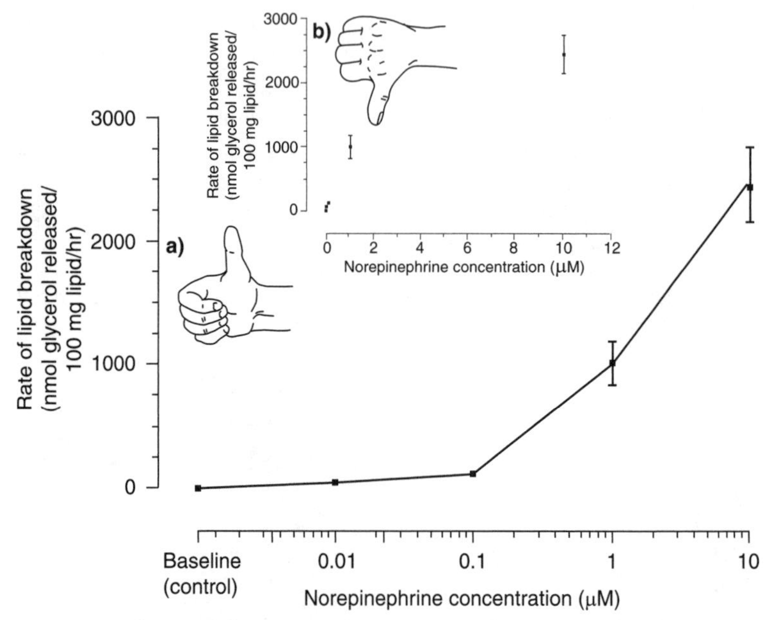

Figure 27. Influence of the neurotransmitter norepinephrine (= noradrenaline) on rate of lipid breakdown in the subcutaneous fatty tissue of hibernating marmots (*Marmota marmota*). Each point represents the mean of 11 determinations, and error bars represent 1 standard deviation about the mean. Note that the norepinephrine concentrations used in the study encompass 4 powers of 10, a situation that typically calls out for logarithmic data representation. (a) The data plotted using a log base 10 scale on the x-axis. (b) (inset) The same data plotted much less effectively, using a normal linear scale.

Based on data of N. Cochet et al., 1999. Regional variation of white adipocyte lipolysis during the annual cycle of the alpine marmot. Comp. Biochem. Physiol. *C 123: 225–232.*

Similarly, compare the clarity of Figure 27*a* with that of the inset (Figure 27*b*). In this case, the data extend through 4 powers of 10. Note how most of the data form a difficult-to-decipher mess in the lower left corner of the inset but are spread out evenly using a scale of log base 10 on the *x*-axis of the main graph.

Although logarithmic graphs are usually plotted using scales in base 10, base 2 scales can also be useful. The hermit crab data of Figure 26*a* are replotted using this log base 2 scale for both axes in Figure 28; now each numbered major tick represents a doubling of crab or shell mass. A quick visual inspection of this graph shows that as the mass of the hermit crab approximately doubles, the mass of the shell it prefers to inhabit also doubles, something not so obvious in the previous plots of the same data (see Fig. 26).

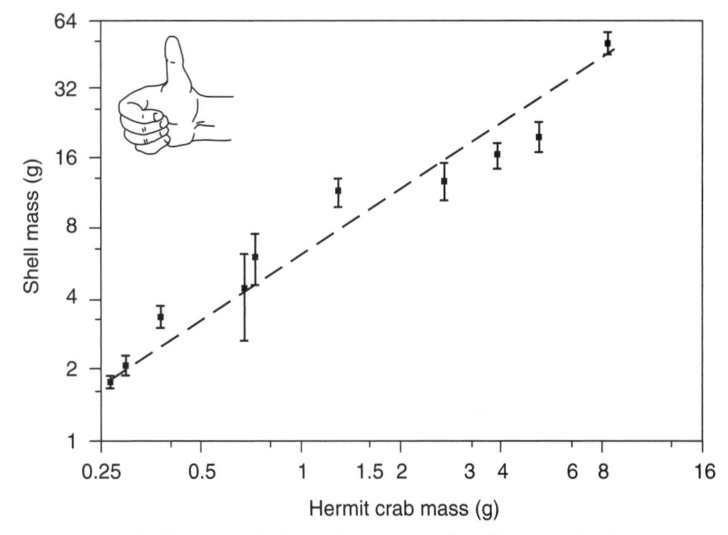

Figure 28. The hermit crab data of Figure 26 plotted using a log base 2 scale on each axis. Such scales are very helpful when you are interested in the effects of doublings, or in how much of a change in x is required to produce a doubling of y.

In some cases, logarithms can also be useful in forcing data into straight lines. Figure 29 shows the relationship between the basal diameter of barnacles (*Balanus amphitrite*) and the dry weight of their tissues. Plotted on a linear scale (Figure 29*a*), tissue weight increases exponentially with increases in barnacle length. Replotting the same data using a double-log scale (base 10), we have essentially bent the data into a straight line (inset) so that the relationship can be represented by a simple equation of the form $y = mx + b$.

Preparing Tables

Tables should always be organized with the data related to a given characteristic being presented vertically rather than horizontally. Tables 6 and 7 present the same information, but in different formats. Table 6 correctly places all information about a single species in one row so that readers can view the information for each species by scanning from left to right and can compare data among different species by scanning up and down a single column. Note also that the independent variable ("species" in this case) is presented vertically in the first column.

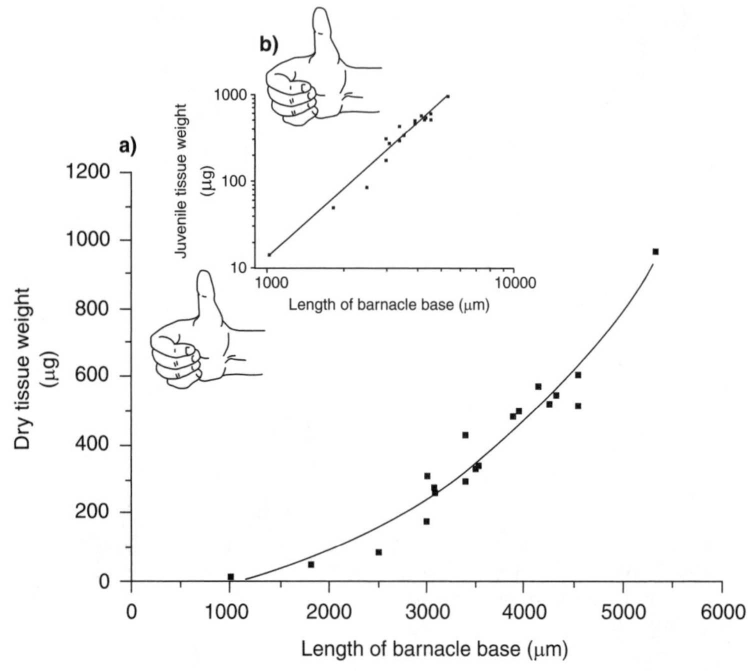

Figure 29. Relationship between barnacle size (in micrometers) and dry tissue weight (micrograms) for juveniles of *Balanus amphitrite.* Each point represents measurements on a single individual. (a) The data plotted on a normal linear scale. (b) The same data replotted using log base 10 scales for both axes. Notice that the exponential curve depicted in (a) becomes a straight line in (b).

Data from Pechenik et al., 1993. Influence of delayed metamorphosis on survival and growth of juvenile barnacles Balanus amphitrite. Mar. Biol. *115: 287–294.*

Table 6. Characteristics of 4 snail populations sampled at Nahant, MA, on October 13, 2011. Data are means of 4 replicate samples ± 1 standard error.

Species	Average Shell Length (cm)	Sample Size	Average No. Animals per m^2
Crepidula fornicata	1.63 ± 0.21	122 indiv.	32.1 ± 4.7
C. plana	1.01 ± 0.34	116	20.8 ± 10.6
Littorina littorea	0.87 ± 0.11	447	113.6 ± 29.1
L. saxatilus	0.40 ± 0.10	60	8.2 ± 5.2

Table 7 is incorrectly organized and so is more difficult to read. Like graphs, tables should be self-sufficient; note how much useful information

Table 7. Characteristics of 4 snail populations sampled at Nahant, MA, on October 13, 2011. Data are means of 4 replicate samples ± 1 standard error.

Species	*Crepidula fornicata*	*C. plana*	*Littorina littorea*	*L. saxatilus*
Av. shell length (cm)	1.63 ± 0.21	1.01 ± 0.34	0.87 ± 0.11	0.40 ± 0.10
Sample size	122 indiv.	116	447	60
Av. no. animals per m^2	32.1 ± 4.7	20.8 ± 10.6	113.6 ± 29.1	8.2 ± 5.2

the author has packed into the legend and column headings of Table 6. Note also that the table legend is placed above the table.

As mentioned earlier (p. xxx), do not present the same data in both a graph and a table: Choose the format that makes it easier for you to get your findings across to readers.

Making Your Figures and Tables Self-Sufficient

A properly executed figure or table is largely self-sufficient: Snipped out of your report along with its accompanying legend or caption, it should make perfectly good sense to any biology major you hand it to. By examining only the axes of a graph and reading the figure caption, for example, the reader should be able to determine the specific question that was asked, how the study was done, and what the main findings are, as discussed in Chapter 3 (pp. 38–43). Consider, for example, Figure 22.

We can tell quite a lot about how and why this study was done just by carefully studying the figure and its caption. The study was apparently undertaken to determine how fast tomato seedlings grow at one particular temperature, 20°C. We know that the growth of 205 seedlings was followed over nearly 2 months, and that seedlings were measured 9 times over that period, approximately once each week. We also know that not all 205 seedlings were measured every time; instead, the researchers subsampled 15–20 each time so that different seedlings were probably measured each time. Finally, we see that each point represents a mean value and that the vertical bars represent 1 standard deviation about the mean.

Similarly, from Figure 18 we know what species was studied and for how long. We also know that caterpillars were maintained at 1 of 5 temperatures, and we know what those temperatures were. Furthermore, we know the number of caterpillars maintained at each temperature.

Try to make your data presentations as self-sufficient as these examples are. In particular, **always indicate the species studied, the sample size** (e.g., Figure 25), **and the number of replicates** (e.g., Table 6). The easier you make it for readers to understand your data, the more likely it is that your work will be read and that you will get your point across intact. Additional advice about constructing good graphs and tables is given in books by W.S. Cleveland and by J. Quinn and M.J. Keough (see Appendix C).

Putting Your Figures and Tables in Order

Once you have your graphs (and other figures like drawings, maps, or photographs) and tables prepared, you need to decide in what order to present them. **Arrange them logically, in the order that you will discuss them.** The first figure that you refer to in the text of your report will be Figure 1, the second will be Figure 2, and so forth. Order your tables in the same way, with Table 1 the first table that you refer to, in order to facilitate a logical presentation of the data.

Incorporating Figures and Tables into Your Report (or Not)

Ask your instructor whether you should incorporate your graphs and tables into the text of the Results section, as they would appear in a published journal article, or whether you should print the figures and tables on separate pages and include them at the end of your report.

To incorporate figures directly into your report, click on the figure to be inserted, and then use the cut and paste function. Alternatively, you can use the Insert function in the Word menu.

Do not insert figures into the text unless your instructor asks you to. Many of us would prefer to see each figure on a separate page at the end of your report; larger figures are much easier to read and comment on.

Verbalizing Results: General Principles

One-sentence Results sections are common in student reports: "The results are shown in the following tables and graphs." However, *common* does not mean *acceptable*. **You must use words to draw the reader's**

attention to the key patterns in your data. But do not simply redraw the graphs in words, as in this description of Figure 20:

> At 20°C, the seedlings showed negligible growth for the first 8 days of study. However, between days 8 and 16, the average seedling grew nearly 5 mm, from about 8 mm to about 13 mm. Growth continued over the next 16 days, with the seedlings reaching an average height of 24 mm by day 24, and 30 mm by day 32.

Let the graph do this work for you; **your task is to summarize the most important findings displayed by the graph** and then to indicate briefly the basis for the statements you make.

First, **decide exactly what you want your reader to see when looking at each graph or table, and then stick the reader's nose right in it**. For example, you might write:

> Temperature had a pronounced effect on seedling growth rates (Fig. 20). In particular, seedlings at 25°C consistently grew more rapidly than those at 20°C.

Remember, in scientific writing every statement of fact or opinion must be backed up with evidence (p. 8). In the example just given, a general statement was supported by reference to a figure, followed by a specific detail that illustrates the point particularly well. In some cases, only a single sentence is required:

> Caterpillars generally fed more slowly on the diet of 10^{-3} M sinigrin than on the wheatgerm controls (Fig. 14).

Readers can then look at Figure 14 and decide whether they see the same trend you do: One sentence and a figure say it all.

Note the use of the past tense in the statement about caterpillar feeding rates:

> Caterpillars generally fed at faster rates on diet *A*.

This statement is quite different from the following one, which uses the present tense:

> Caterpillars feed at faster rates on diet *A*.

By using the present tense, you would be making a broad generalization extending to all caterpillars, or at least to all caterpillars of the species tested. Before one can make such a broad statement, the experiment must be repeated many, many times, and similar results must be obtained each time. After all, the writer is making a statement about all caterpillars under all conditions. By sticking with the past tense here, you are clearly referring only to the results of your study. Be cautious: **Always present your results in the past tense.**

Note also that the authors of these examples **do not make the reader interpret the data**. You must tell your readers exactly what you want them to see when they look at your table or graph. Consider the following 2 examples. The data concern the shell lengths of a particular snail species collected along a rocky coastline from 2 regions between the high-tide level and the low-tide level. I have not included any figures or tables from this survey, so I refer to them only as "xx" in these examples:

Although individual specimens of *Littorina littorea* varied considerably in shell length at each tidal height (Fig. xx), there was a significant ($t = 26.3$; $d.f. = 47$; $p < 0.05$) distributional effect of shore position on mean size (Table xx).

Although individual specimens of *Littorina littorea* varied considerably in shell length at each tidal height (Fig. xx), the mean shell length was significantly greater ($t = 26.3$; $d.f. = 47$; $p < 0.05$) for snails collected higher up in the intertidal zone (Table xx); high shore animals were, on average, 26% larger than low-shore animals.

In the first of these examples, the author expects the readers to figure out what specific information is important in the data. Sometimes this is because the author has not taken the time to think carefully enough about the data. The modified version conveys quite a different impression: We know exactly what the author wants us to see, and we can then decide whether we agree with the author's statements.

Note that in the model example just presented, the lead factual statement was supported both by reference to a figure and by the results of statistical analyses. **Note also that the author provided not just the *p*-value associated with the result (to indicate how convincing the difference among means is) but also the name of the test statistic (a *t*-statistic in this example, resulting from a *t*-test) and the number of degrees of freedom associated with the analysis** (d.f., related to sample size as discussed in Chapter 4). This allows readers to check the validity of the statistical test performed and the validity of the results obtained. Also note that the author correctly refers to the observed differences in shell length as being "significant." **Saying that differences are "significant" (or "not significant") implies that you have subjected your data to rigorous statistical testing.** If you have not conducted statistical analyses, it is perfectly fair to write "Temperature had a pronounced effect on seedling growth rates," or "Seedlings treated with nutrients appeared to grow at slightly faster rates than those treated with distilled water," referring readers to the appropriate table or figure. But if you cannot provide statistical support, you *cannot* say that seedlings in one treatment grew *significantly* faster than those in another treatment. Note also that the author of this example indicates the magnitude of the effect observed; the size differences are not just statistically significant but are, in fact, quite large (26%).

As discussed more fully in Chapter 4, even if you analyze your data statistically, you must nevertheless be cautious in drawing conclusions. Your data may support one hypothesis more than another, but they cannot *prove* that any hypothesis is true or false. Chapter 4 is worth reading even if you are not required to conduct statistical analyses of your data; as biologists in training, the *why* of statistical analysis is more important to you than the *how*.

Verbalizing Results: Turning Principles into Action

Let us apply these principles to the caterpillar study discussed earlier in this chapter. First, is there anything about the general response of the animals worth drawing attention to? You might, for example, be able to write:

> All the caterpillars were observed to eat throughout the experiment.

More likely, living things behaving as they do, you will say something like:

> One of the animals offered diet *A* and 2 of the animals offered diet *B* were not observed to eat during the 3 h experiment, and the results from these animals were therefore excluded from analysis.

Such a decision to exclude data from further analysis is fine, by the way—it won't be considered data falsification—as long as you indicate the reason for the decision, and as long as the decision is made objectively; as discussed earlier, **you cannot exclude data simply because they violate a trend that would otherwise be apparent or because the data contradict a favored hypothesis**.

Next, **go back to your initial list, and reword each question as a statement**. For example, the first question posed on p. 166 ("Did the caterpillars feed at different rates on the different diets?") might be reworded as

> Caterpillars generally fed at slower rates on the 10^{-3} M sinigrin diet than on the wheatgerm controls (Fig. 14).

or

> Caterpillars fed at significantly slower rates on the 10^{-3} M sinigrin diet than on either the lower-concentration sinigrin diet or the wheatgerm controls (Fig. 14) ($F = 30.3$, $d.f. = 2, 11$; $p < 0.0001$).

If you follow this procedure for each question on your list, your Results section will be complete. The written part will generally be quite short.

Writing about Negative Results

An experiment that was correctly performed always "works." The results may not be what you had expected or hoped for, but this does not mean that the experiment has been a waste of time. If biologists threw away their data every time something unexpected happened, we would rarely learn anything new. The data you collect are real; only the interpretation of those data is open to question. Therefore, always treat your data with respect. The lack of a trend, or the presence of a trend contrary to expectation, is itself a story worth telling. See also pp. 66 through 69 for related advice about dealing with statistical analyses that fail to support a favored hypothesis or expected outcome.

Writing about Numbers

According to the Council of Science Editors (*Scientific Style and Format: The CSE Manual for Authors, Editors, and Publishers,* 7th edition), you should **use numerals rather than words when writing about counted or measured items, percentages, decimals, magnifications, and abbreviated units of measurement** (see pp. 64–67): 6 larvae, 18 seedlings, 25 drops, 25%, 1.5 times greater, 50×· magnification, a 3:1 ratio, 0.7 g, 18 ml.

All rules, however, have exceptions. Use words rather than numerals if beginning a sentence with a number or percentage:

> Twenty grams of NaCl were added to each of 4 flasks.
>
> Thirty percent of the tadpoles metamorphosed by the end of the second week.

Or, avoid starting sentences with numbers, as in this rewrite of the first example:

> To each of 4 flasks we added 20 g of NaCl.

When 2 numbers are written adjacent to each other without being separated by words or a comma, write one of the numbers in words:

> The sample was divided into five 25-seedling groups.

Better still, that sentence could easily be rewritten so that numerals are appropriate for both numbers:

> The samples were divided into 5 groups of 25 seedlings each.

Zero and one present special problems: a "0" is easily mistaken for the letter "O," and the numeral "1" is easily mistaken for the letters "l" (as in the abbreviation for "liters") or "I." Plus, it just looks odd to read, "I know that 1 day my prince will come." The Council of Science Editors has recently clarified its position on these issues. In most cases, unless the zero or one is followed by units of measure (e.g., "I added 1 mg of sucrose to the solution"), or is part of an equation (e.g., "$n = 1$"), or is part of a series that includes larger numbers ("1, 8, and 25 individuals... "), use words rather than numerals for zero and one.

When writing about numbers smaller than zero, precede the decimal point with a zero:

> ... and we then added 0.25 g NaCl to each flask.

When using ordinal numbers (e.g., first, fifth), the Council of Science Editors suggests using words for the first 9 numbers and numerals for the others ("the 25th replicate..."). However, you should be consistent within a series ("first, fifth, ninth" and "13th, 14th, 15th"; but "5th, 9th, 15th," rather than "fifth, ninth, 15th").

When writing about very large or very small numbers, particularly in association with concentrations or rates, **use scientific notation**. It is preferable, for example, to write about solute concentrations of $5.6 \cdot 10^{-3}$ g/ml rather than 0.0056 g/ml, and about cell concentrations being approximately $1.8 \cdot 10^{5}$ cells/ml rather than about 180,000 cells/ml. Note that I could have avoided both scientific notation and commas in the first example by describing the solute concentration as 5.6 mg/ml. By the way, the word *per*, as in "cells per ml" or "distance per second," may be indicated using either a slash or an exponent: "$1.8 \cdot 10^{5}$ cells/ml" and "$1.8 \cdot 10^{5}$ cells ml^{-1}" are equally acceptable forms of expression. See Technology Tip 1 (p. 18) to learn about programming Word to produce such formatting automatically.

Finally, the Council of Science Editors recommends **using commas** only when numerals contain 5 or more numbers, as in the following example:

> Only 1073 of the original 12,450 frog tadpoles died during the study.

In Anticipation—Preparing in Advance for Data Collection

Much of the work involved in putting together a good research report goes into preparing the Results section. You can save yourself considerable effort and frustration by planning ahead before you enter the laboratory to do the experiment. Be prepared to record your data in a format that will enable you to make your calculations easily. For the caterpillar experiment referred to previously, come to the laboratory with a data sheet set up like the one shown in Figure 30. Using this data sheet, the data are recorded in the *x* areas during the laboratory period; the blank spaces will be filled in later, as you make your calculations. If possible, leave a few blank columns at the right to accommodate unanticipated needs discovered as you record or work up your data.

Figure 30. Sample format for a laboratory data sheet.

In introductory laboratory exercises, students are often provided with data sheets already set up in a useful format. Take a careful look at those data sheets to understand how they are organized and why they appear as they do; in more advanced laboratory courses, you will be responsible for organizing your own data sheets. As mentioned earlier, **always follow any number you write down with the appropriate units**, such as mg (milligrams), cm (centimeters), or mm/min (millimeters per minute, often written as mm min^{-1}). This will avoid potential confusion later.

You can also enter your data directly into a computer if you set up a data sheet in advance (see Technology Tip 6).

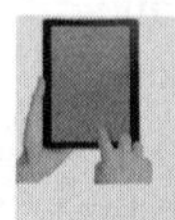

TECHNOLOGY TIP 6

Using computer spreadsheets for data collection

One advantage of entering data directly into a spreadsheet (e.g., Excel) is that you can preprogram automatic calculations into the spreadsheet. Consider, for example, Table 3 (p. xxx). If we enter data into columns 2, 3, and 5 (Initial Caterpillar Wt., Final

Caterpillar Wt., and Wt. of Food Lost), we can enter formulas into columns 4 and 6 so that Caterpillar Wt. Change and Feeding Rate will be calculated automatically. Before entering formulas in Excel, remember to first click on the appropriate box and then click "=" to indicate that a formula is on its way.

Whenever you enter formulas into a spreadsheet, make one sample calculation by hand or with a calculator to be sure that the formula you entered is correct. Also check to be sure that the formula is correctly applied to all entries in each column. Especially useful operators include SUM, AVERAGE, and STDEV. For example "= AVERAGE(B3:B21)" calculates the average of the numbers in columns B3–B21 of the spreadsheet, whereas "= STDEV(B3:B21)" returns the standard deviation about the mean.

Following the advice of the previous paragraph can save you hours of work later on. Even so, it takes time and care to put together an effective Results section. But this section is the heart of your report. Craft it properly, and the remainder of the work will be relatively easy.

CITING SOURCES

The next sections to prepare are the Discussion section and the Introduction. In both sections, you will be making statements of fact that require support, often from written sources. As stated in Chapter 1 (Rule 8, p. 8 and discussed more fully in Chapter 5, **every statement of fact or opinion must be supported with a reference to its source**. Read Chapter 5 (pp. 69–74) for specific instructions on citing sources.

WHAT TO DO NEXT?

I usually draft my Discussion section next, although you might instead try drafting your Introduction (p. 203). See what works best for you. The key is to undertake neither one until you have your Materials and Methods section and your Results section in good order.

WRITING THE DISCUSSION SECTION

> *The scientific method does not require researchers to be unbiased observers of nature. Scientists almost always have a theory in mind when they perform an experiment. But the method does require that scientists be willing to change their views about nature when the data demand it.*
>
> R.M. HAZEN AND J. TREFIL

In the Discussion section of the report, you must interpret your results in the context of the specific questions you set out to address during your experiment and in the context of any relevant broader issues that have been raised in lectures, textbook readings, previous coursework, and possibly your library research. You will consider the following issues:

1. What did you expect to find, and why?
2. How did your actual results compare with the results you expected? If you set out to test specific hypotheses, do your data support one hypothesis more than another or allow you to eliminate one or more of them? Explain your logic.
3. How might you explain any unexpected results?
4. How might you test those potential explanations?
5. Based on your results, what question or questions might you logically want to ask next?

Clearly, if your results coincide exactly with those expected from prior knowledge, your Discussion section will be rather short, but such a high level of agreement is rarely obtained in 3- or 4-hour laboratory studies. Indeed, high degrees of variability characterize many aspects of research in biology, especially at the level of the whole organism. After all, genetically based variation in traits is the raw material of evolution: Without such variation, evolution by natural selection would not be possible. Often a study will need to be repeated many times, with very large sample sizes, before convincing trends emerge. This point is discussed further in Chapter 4, beginning on page 60. A short paper in a biological journal may well represent years of work by several competent, hardworking individuals. Even the simplest of questions is often not easily answered. Nevertheless, every experiment that was carried out properly tells you *something*, even if that something is not what you specifically intended to find out.

Expectations

State your expectations explicitly, and back up your statements with a reference. Scientific hypotheses are not simply random guesses. Your expectations must be based on facts, not opinions; these facts could come from lectures, laboratory manuals or handouts, textbooks, or any other traceable source. In discussing a study on the effectiveness of different wavelengths of light in promoting photosynthesis, for example, you might write something like the following:

> All wavelengths of light are not equally effective in promoting photosynthesis: green light is said to be especially ineffective (Ellmore and Mirkin, 2009). This is because green light tends to be reflected rather than absorbed by plant pigments, which is why most plants look green (Ellmore and Chew, 2003). Our results supported this expectation. In particular...

Alternatively, a Discussion section might profitably begin as follows:

> The results of our experiment failed to support the hypothesis (McClaughlin and McVey, 2001) that caterpillars of *Manduca sexta* reared on one uniquely flavored diet will prefer that diet when subsequently given a choice of foods.

Here we have managed to state our expectations and compare them with our results in a single sentence. In both cases, we have begun our discussion on firm ground—with facts rather than unsupported opinions.

Note that in this last example, the expectations were based on previously published research. Your expectations might instead be based on a hypothesis stated in your Introduction and tested in the Results section.

Explaining Unexpected Results

When results refuse to meet expectations, students commonly blame the equipment, the laboratory instructor, their laboratory partners, or themselves. Generally, more scientifically interesting possibilities than experimenter incompetence are the culprits. A few years ago, I did an experiment with 2 colleagues to see if dopamine receptors were involved in the pathway that led to metamorphosis in a particular marine animal. The idea was to add to seawater a chemical known to disable dopamine receptors and then see if the larvae could nevertheless be made to metamorphose. Every time we did the experiment, we got a different result.

Sometimes the chemical blocked metamorphosis, but in other experiments it actually stimulated larvae to metamorphose! What on earth was going on? It turned out that the response of competent larvae to the chemical changes predictably as larvae age—something never reported before. The paper we ended up publishing* was very different from the paper we had originally envisioned. Yes, dopamine receptors are involved in the metamorphic pathway of this species, but this turned out to be, in my view, the least interesting aspect of the work.

Don't be too hard on yourself if your results don't fit your expectations, or if they don't disprove your null hypothesis (Chapter 4) when you expected them to. **Base your discussion on the data you obtain.** And don't limit yourself to assessing *a priori* hypotheses: Probe your data thoroughly, expect the unexpected, and consider all potential aspects of, and reasonable explanations for, your data. Take another look at the list of factors you wrote when beginning to work on your Materials and Methods section. Could any of these factors be sufficiently different from the normal or standard conditions under which the experiment is performed to account for the difference in results? Look again at your laboratory manual or handout. Are any of the conditions under which your experiment was performed substantially different from those assumed in the instruction manual? If you discover no obvious differences in the experimental conditions, or if the differences cannot account for your results, include this point in your report, as in this example:

> The discrepancy in results cannot be explained by the unusually low temperature in the laboratory on the day of the experiment, since the control animals were subjected to the same conditions and yet behaved as expected.

If potentially important differences are noted, put this ammunition to good use:

> In prior years, these experiments have been performed using species *X* (Professor E. Iyengar, personal communication). It is possible that species *Y* simply behaves differently under the same experimental conditions.

Note that the writer does not *conclude* that species *X* and species *Y* behave differently; the writer merely *suggests* this explanation as a possibility.

---

*Pechenik, J.A., Cochrane, D.E., Li, W. 2002. Timing is everything: The effects of putative dopamine antagonists on metamorphosis vary with larval age and experimental duration in the prosobranch gastropod *Crepidula fornicata*. *Biol. Bull.* 202: 137–147.

Always be careful to distinguish possibility from fact. Suggesting a logical possibility won't get you into any trouble. Stating your idea as though it was an accepted fact, on the other hand, is sticking your neck out far enough to get your head chopped off.

Continue your discussion by indicating possible ways that the differences in behavioral responses **might be tested**. For example:

> This possibility can be examined by simultaneously exposing individuals of both species to the same experimental conditions. If species *X* behaves as expected and species *Y* behaves as it did in our experiment, then the hypothesis of species-specific behavioral differences will be supported. If species *X* and species *Y* both respond as species *Y* did in the present study, then some other explanation will be called for.

Continue in this vein, evaluating all the reasonable, testable possibilities you can think of. An instructor enjoys reading these sorts of analyses, because they indicate that students have been thinking about what they've done.

Notice that in the preceding example, the writer did not say, "If species *X* behaves as expected and species *Y* behaves as it did in the present experiment, then the hypothesis will be supported." This writer remembers rule number 14 (p. 9: **Never make the reader back up.** Notice, too, that the writer did not say, "then the hypothesis will be true," or "then the hypothesis will be proven." **Experiments cannot *prove* anything; they can only support or not support specific hypotheses.** As scientists, our interpretations of phenomena may make excellent sense based on what we know at the moment, but those interpretations are not necessarily correct. New information often changes our interpretations of previously acquired data.

Analysis of Specific Examples

Example 1

In this study, tobacco hornworm caterpillars were raised for 4 days on one diet, and then tested over a 3-hour period to see if they preferred that food when given a choice of diets.

Student Presentation

> The data indicate that the choice of food was not related to the food upon which the caterpillars had been reared. These data run counter to the hypothesis (Back and Reese, 1976) that

hornworms are conditioned to respond to certain specific foods. Only 1 set of data out of the 4 gave any indication of a preference for the original diet, and that indication was rather weak.

There are many possible explanations for data that are so contrary to previous experimental results. Inexperience of the experimenters, combined with the fact that 3 different people were recording data about the caterpillars, may account for part of the error. Keeping track of many worms and attempting to interpret their actions as having chosen a food or having merely been passing by may have proven to be too much for first-time hornworm watchers. The mere fact that each of 3 people will interpret actions differently and will have somewhat different methods of recording information introduces bias into the data.

Analysis

This Discussion section starts out well, with a comparison between the results expected and the results obtained. The hypothesis being discussed is clearly stated, and a supporting reference is given. The student even recognizes that "data are" rather than "data is." The second paragraph, however, betrays a total lack of confidence in the data obtained; the results could not possibly have turned out this way unless the researchers were incompetent, writes the student. Although inexperience can certainly contribute to suspicious results, are there no other possible explanations? Does it really take years of training to determine whether a caterpillar ate food *A* or food *B*?

Compare this report with the one in the next example. This Discussion section deals with the same experiment. In fact, the 2 students were laboratory partners.

Example 2

Contrary to expectation, our results suggest that caterpillars of this species showed no preference in the diet they touched first and the diet they spent the most time feeding on. This unexpected finding may be due to the fact that the caterpillars were not reared on the original diets long enough to acquire a lasting feeding preference. They were reared on the original diets for only 4 days, whereas the laboratory handout had suggested a pre-feeding period of 5–10 days (Orians and Starks, 2003). This possibility may be tested by performing the same experiment

> but varying the amount of time that the caterpillars are reared on the original diets. Such an experiment would determine whether there is a critical time that caterpillars should be reared on a particular diet before they will show a preference for that diet. Another possible explanation for our results is that the caterpillars used in our study were very young, weighing only 3–6 mg. Finally, this experiment lasted only 3 hours. Perhaps different results would have been obtained had the organisms been given more time to adjust to the test conditions. We suggest conducting an identical experiment for a longer period of time, such as 10–12 hours.

The author of this report produced a paper that clearly indicates thought. Which report do you suppose received the higher grade?

Example 3

In this experiment, several hundred milliliters (ml) of filtered pond water were inoculated with a small population of the ciliated protozoan *Paramecium multimicronucleatum* and then distributed among 3 small flasks. Over the next 5 days, changes in the numbers of individuals per ml of water in each flask were monitored.

Student Presentation

> The large variation observed between the groups of 3 replicate populations suggests that the experimental technique was imperfect. The sampling error was high because it was difficult to be precise in counting the numbers of individuals. Some animals may have been missed while others were counted repeatedly. More accurate data may be obtained if the number of samples taken is increased, especially at the higher population densities. In addition, more than 3 replicate populations of each treatment could be established. Finally, extremely precise microscopes and pipets could be used by experienced operators to reduce sampling error.

Analysis

This writer, like the writer of Example 1, starts out by assuming that the experiment was a failure and then spends the rest of the report making excuses for this failure. The quality of the microscopes was certainly adequate to recognize moving objects, and *P. multimicronucleatum* was the only organism moving in the water: The author is grasping at straws. If the author had more confidence in his or her abilities, the paper might have been far different. Isn't there some chance that the experiment was performed correctly? Lacking confidence in the data, the student took the easy way out and looked no further, even though he or she actually had access to data that would have allowed several of the stated hypotheses to be assessed. On each day, for example, several sets of samples were taken from each flask, and each set gave similar estimates for the numbers of organisms per milliliter: This consistency of results suggests that the variation in population density from flask to flask was real and not due to experimenter incompetence. In addition, although the student stated correctly that larger sample sizes would have been helpful, he or she should have supported that statement with additional data analysis. Fifteen drops were sampled from each flask for each set of samples. The student could have calculated the mean number of individuals in the first 3 drops, the first 6 drops, the first 9 drops, the first 12 drops, and then the full 15 drops, to see how the estimates of population size changed as the sample size increased. With such calculations, the student would probably have found that larger sample sizes are especially important when population density is low. (Why might this be so?)

Example 4

In this last study, a group of students went seining for fish in a local pond. Every fish was then identified to species. It turned out that 91% of the fish in the sample of 73 individuals belonged to a single species. The remaining fish were distributed among only 2 additional species.

Student Presentation

I find the small number of species represented in our sample surprising, since the pond is fed by several streams that might be expected to introduce a variety of different species into it, assuming that the streams are not polluted. The lab manual states that 12 fish species have been found in the adjacent streams. It appears that the conditions in the pond at the time of our sampling were especially suitable for one species in particular out of all those that most likely have access to it. Perhaps the physical nature of the pond is such that the number of niches is small, in which case competition would become very keen; only one species can occupy a given niche at any one time (Ricklefs and Miller, 2000). The reproductive pattern of the fishes might also contribute to the observed results. Possibly *Lepomis macrochirus*, the dominant species, lays more eggs than the others, or perhaps the young of this species survive better, or prey on the young of other species.

Another possible explanation for our findings is that we sampled only the perimeter of the pond, since our seining was limited to a depth of water not exceeding the height of the seiners. The species distribution could be very different in the middle of the pond at a greater depth.

Analysis

I have not reproduced the entire Discussion section of the student's paper, but even this excerpt demonstrates that a little thinking goes a long way. Note that the student did not require much specialized knowledge to write this Discussion section, only a bit of confidence in the data. Another student might well have written:

> Most likely, the fish were incorrectly identified; more species were probably present than could be recognized by our inexperienced team. It is also possible that the net had a large tear, which let the members of other species escape. I didn't notice this rip in the fabric, but my glasses were probably dirty, and then again, I'm not very observant.

WRITING THE INTRODUCTION SECTION

The Introduction section establishes the framework for the entire report. In this section, **you briefly present background information that leads to a clear statement of the specific issue or issues that will be addressed in the remainder of the report**; by the time you have finished writing the Materials and Methods, Results, and Discussion sections of your laboratory report, you should be in a good position to know what these issues are. In 1 or 2 paragraphs, then, you must present an argument explaining why the study was undertaken. More to the point, perhaps, the Introduction provides you with your first opportunity to convince your instructor that you understand why you have been asked to do the exercise.

Every topic that appears in later sections of your report should be anticipated clearly in the Introduction, and the Introduction should contain only information that is directly relevant to the rest of the report.

Stating the Question

Even though the statement of questions posed or of issues addressed generally concludes the Introduction section of a report, it is useful to deal with this issue first. What was the *specific* issue or question addressed in your study?

First, write the following words: "In this study" or "In this experiment." Then complete the sentence as specifically as possible. Three examples follow:

In this study, the oxygen consumption of mice and rats was measured to investigate the relationships between metabolic rate, body weight, and body surface area.

In this study, we collected fish from 2 local ponds and classified each fish into its proper taxonomic category.

In this experiment, we asked the following question: Do the larvae of *Manduca sexta* prefer the diet upon which they have been reared when offered a choice of diets?

Note that each statement of intent is phrased in the past tense since the students are describing studies that have now been completed.

The strong points of these statements are best revealed by examining a few unsatisfactory alternatives:

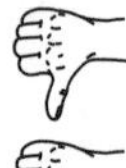

In this study, we measured the metabolic rate of rats and mice.

In this study, we worked with freshwater fish.

In this experiment, the feeding habits of *Manduca sexta* larvae were studied.

Each of these 3 unsatisfactory statements is vague; readers will assume, perhaps correctly, that you are as much in the dark about what you've done as your writing implies. **Be specific.** Here, in one sentence, you must come fully to grips with the goals of your study. There *was* some point to the time that you were asked to spend in the laboratory; find it.

If you go on to state specific expectations or to present specific hypotheses that you set out to test, **make the basis for those hypotheses clear**, as in the following example:

We expected larger mice to respire faster than smaller mice, since larger mice support a greater biomass. However, we also predicted that respiration rates per gram of tissue would be similar in large

> and small individuals, since dividing by weight should adjust for size differences.

As mentioned earlier (p. 197), you must provide a rationale (e.g., logical argument or results from prior studies) for any expectations or specific hypotheses that you state.

An Aside: Studies versus Experiments

An experiment always involves manipulating something, such as an organism, an enzyme, or the environment, in a way that will permit specific relationships to be examined or hypotheses to be tested. Containers of protozoans in pond water could be distributed among 3 temperatures, for example, to test the influence of temperature on the reproductive rate of the particular species under study. The ability of salivary amylase to function over a range of pHs might be examined to test the hypothesis that the activity of this enzyme is pH sensitive. In the field, a population of marine snails from one location might be transplanted to another location and the subsequent survival and growth of the transplanted population studied so as to test the hypothesis that conditions in the new location are less hospitable for that species than in the location from which the original population was obtained. As a control, of course, the survival and growth of animals not transplanted would also have to be monitored over the same period. Note that an experiment may be conducted in the laboratory or in the field.

It is permissible to refer to experiments as "studies," but not all studies are "experiments." In contrast to the preceding experiments, some exercises require you to collect, observe, enumerate, or describe. You should avoid referring to such studies as experiments; **when there are no manipulations, there are no experiments**. You might, for example, collect insects from light fixtures located at several different locations within the biology building and identify them to the species level, enabling you to examine the distribution of insect species within the building. Or you might be asked to provide a detailed description of the feeding activities of an insect. Or you might spend an afternoon documenting the depth to which light penetrates in various areas of a lake and then correlate that information with data on the distribution of aquatic plants in the different areas. In each case, you should refer to your work as a study, not as an experiment. For example:

> In this study, insects were collected from all light fixtures on floors 1, 3, and 5 of the Dana building, and the distribution of species among the different locations was determined.

Providing the Background

Having posed, in a single sentence, the question or issue that was addressed, it will now be easier to fill in the background needed to understand why the question was asked. A few general rules should be kept in mind:

1. **Support all statements of fact with a reference to your textbook, laboratory manual, outside reading, or lecture notes**. Unless you are told otherwise by your instructor, do not use footnotes. Rather, refer to your reference within the text, giving the author of the source and the year of publication, as in the following example:

    > Many marine gastropods enclose their fertilized eggs within complex encapsulating structures (Hunt, 1966; Tamarin and Carriker, 1968).

    Note that the period concluding the sentence comes after the closing parenthesis. Do not cite any material that you have not read.
2. **Define specialized terminology.** Your instructor probably knows the meaning of the terms you will use in your report, but by defining them in your own words, you can convince the instructor that you, too, know what these words mean. Write to illuminate, not to impress. As always, if you write with your future self in mind as the audience, you will usually come out on top; write an Introduction you will be able to understand 5 years from now. The following examples obey this and the preceding rule:

    > A number of caterpillar species are known to exhibit induction of preference, a phenomenon in which an organism develops a preference for the particular flavor on which it has been reared (Westneat and Sesterhenn, 1983).

    > The development of mature female gametes, a process termed oogenesis, is regulated by changing hormonal levels in the blood (Gilbert, 2006; McVey, 2008).
3. **Never set out to prove, verify, or demonstrate the truth of something.** Rather, set out to test, document, or describe. In biology (and science in general), truth is elusive; it is important to keep an open mind when you begin a study and when you write up the results of that study. It is not uncommon to repeat someone else's experiment or observations and obtain a different result or description. Responses will differ with species; time of year;

and other, often subtle, changes in the conditions under which the study is conducted. To show that you had an open mind when you undertook your study, you would want to revise the following sentences before submitting them to your instructor:

In this experiment, we attempted to demonstrate induction of preference in larvae of *Manduca sexta.*

This experiment was designed to show that pepsin, an enzyme promoting protein degradation in the vertebrate stomach, functions best at a pH of 2, as commonly reported (Bernheim and Cochrane, 1999).

The first example might be modified to read:

In this experiment, we tested the hypothesis that young caterpillars of *Manduca sexta* demonstrate the phenomenon of induction of preference.

4. **Be brief.** The Introduction must not be a series of interesting but random facts about the topic you have investigated. Rather, **every sentence should be designed to directly prepare the reader for the statement of intent**, which will appear at the end of the Introduction section, as already discussed. If, for example, your study was undertaken to determine which wavelengths of light are most effective in promoting photosynthesis, there is no need to describe the detailed biochemical reactions that characterize photosynthesis. As another example of what not to do, consider these few sentences taken from a report describing an induction-of-preference study. Caterpillars were reared on one diet for 5 days and tested later to see if they chose that food over foods that the caterpillars had never before experienced.

In this experiment, we explored the possibility that larvae of *Manduca sexta* could be induced to prefer a particular diet when later offered a choice of diets. The results of this experiment are important because induction of preference is apparently linked to (1) the release of electrophysiological signals by sensory cells in the animal's mouth and (2) the release of particular enzymes, produced during the period of induction, that facilitate the digestion and metabolism of secondary plant compounds (laboratory handout, 2010).

The entire last sentence does not belong in the Introduction. The work referred to in this example was a simple behavioral study: Students did not make electrophysiological recordings, nor did they isolate and characterize any enzymes. Although a consideration of these 2 topics might profitably be incorporated into a discussion of the results obtained, these issues should be excluded from the Introduction, because they do not explain why this particular study was undertaken. **Include in your Introduction section only information that prepares the reader for the final statement of intent.** You might, on a separate piece of paper, jot down other ideas that occur to you for possible use in revising your Discussion section, but if they don't contribute to your Introduction, don't let them intrude. Be firm. Stay focused.

5. **Write an Introduction for the study that you ended up doing.** Sometimes it is necessary to modify a study for a particular set of conditions, with the result that the observations actually made no longer relate to the questions originally posed in your laboratory handout or laboratory manual. For example, the pH meter might not have been working on the day of your laboratory experience, and your instructor modified the experiment accordingly. Perhaps the experiment you actually performed dealt with the influence of temperature, rather than pH, on enzymatic reaction rates. In such an instance, you would not mention pH in your Introduction section since the work you ended up doing dealt only with the effects of temperature.
6. Once you have written a first draft of your Introduction, idea mapping (pp. 84–88) can help you decide which ideas to keep and help you to organize those ideas to best advantage.

A Sample Introduction

The following paragraphs satisfy all the requirements of a valid Introduction. This Introduction section is brief but complete—and effective:

> It is well known that plants can use sunlight as an energy source for carbon fixation (Ellmore and Reed, 1993). However, all wavelengths of light need not be equally effective in promoting such photosynthesis. Indeed, the green coloration of most leaves suggests that wavelengths of approximately 550 nm are reflected rather than absorbed so that this wavelength would not be expected to produce much carbon fixation by green plants.
>
> During photosynthesis, oxygen is liberated in proportion to the rate at which carbon dioxide is fixed (Ellmore and Reed, 1993).

> Thus, relative rates of photosynthesis can be determined either by monitoring rates of oxygen production or by monitoring rates of carbon dioxide uptake. In this experiment, we monitored rates of oxygen production to test the hypothesis that wavelengths of light differ in their ability to promote carbon fixation by the aquatic plant *Elodea canadensis.*

Note how, in this Introduction, the material progressed from a rather general statement (plants photosynthesize) to more specific statements and, finally, to the specific research objectives of the study. You will see the same progression in the Introduction sections of most published studies. Your instructor should also see it in yours. Note also that every sentence leads the reader logically to the final sentence.

TALKING ABOUT YOUR STUDY ORGANISM OR FIELD SITE

If your study organism or field site was deliberately chosen because it was ideally suited to investigating the particular problem that you addressed, conclude your Introduction with a brief paragraph explaining your choice. Otherwise, that information would be more appropriate in your Materials and Methods section, as on p. 162. Here is an example of how the material presented on p. 162 could be rewritten as a fine ending to an Introduction:

> *Hydroides dianthus* is an excellent organism for such a study, as its larvae can be obtained in great numbers almost year-round and reared in the laboratory with greater than 90% survival (Qian, 2000; Toonen and Pawlik, 2001). Moreover, the larvae become capable of metamorphosing within 4–6 days at 25°C (Scheltema, 1981; Bryan and Qian, 1997) and can be readily induced to metamorphose by simply elevating the potassium concentration of seawater by 15 mM* (Bryan and Qian, 1997).

DECIDING ON A TITLE

A good title summarizes, as specifically as possible, what lies within the Introduction and Results sections of the report. For this reason,

---

*millimolar = 10^{-3} moles per liter.

write your title after you have written the rest of your report. Your instructor is a captive audience. In the real world of publications, however, your article will vie for attention with many other articles written by many other people; the busy potential reader of your paper will glance at the title of your report and promptly decide whether to stay or move on. **The more revealing your title, the more easily potential readers can assess the relevance of your paper to their interests.** A paper that delivers something other than what is promised by the title can lose you considerable goodwill when read by the wrong audience, and it may be overlooked by the very readers for whom the paper was intended. Indeed, many potential readers will miss your paper entirely, because indexing services, such as *Biological Abstracts* and *ISI Web of Science,* use keywords from a paper's title in preparing their subject indexes.

Here is a list of mediocre titles, each followed by 1 or 2 more revealing counterparts:

1. **No:** Metabolic rate determinations

   Yes: Exploring the relationship between body size and oxygen consumption in mice

2. **No:** The role of a homeobox gene

   Yes: The homeobox gene *Irx5* is needed for retinal cell development in mice

3. **No:**

   a. Measuring the feeding behavior of caterpillars

   b. Food preferences of *Manduca sexta* larvae

   Yes:

   a. Measurements of feeding preferences in tobacco hornworm larvae (*Manduca sexta*) reared on 3 diets

   b. Can larvae of *Manduca sexta* (Arthropoda: Insecta) be induced to prefer a particular diet?

4. **No:** Effects of pollutants on sea urchin development

   Yes: Influence of Cu^{2+} on fertilization success and gastrulation in the sea urchin *Strongylocentrotus purpuratus*

The original titles are too vague to be compelling. Why go out of your way to give potentially interested readers an excuse to ignore your paper? Of more immediate concern in writing up laboratory reports rather than journal articles is this suggestion: Why not use a title that demonstrates to your instructor that you have understood the point of the exercise? Win your reader's confidence right at the start of your report. (By the way, the title should appear on a separate page, along with your name and the date on which your report is submitted.)

WRITING AN ABSTRACT

The Abstract, if requested by your instructor, is placed at the beginning of your report, immediately following the title page. Yet it should be the last thing that you write, other than the title, since it must completely summarize the entire report: why the experiment was undertaken, what problem was addressed, how the problem was approached, what major results were found, and what major conclusions were drawn. And it should do all this in a single paragraph.

Despite its unimpressive length, a successful abstract is notoriously difficult to write. **In compact form, your abstract must present a complete and accurate summary of your work, and that summary must be fully self-contained**—that is, it must make perfect sense to someone who has not read any other part of your report, as in the following example:

> This study was undertaken to determine the wavelengths of light that are most effective in promoting photosynthesis in the aquatic plant *Elodea canadensis* since some wavelengths are generally more effective than others. Rate of photosynthesis was determined at 25°C, using wavelengths of 400, 450, 500, 550, 600, 650, and 700 nm and measuring the rate of oxygen production for 1-h periods at each wavelength. Oxygen production was estimated from the rate of bubble production by the submerged plant. We tested 4 plants at each wavelength. The rate of oxygen production at 450 nm (approximately 2.5 ml O_2/mg wet weight of plant/h) was nearly 1.5· greater than that at any other wavelength tested, suggesting that light of this wavelength (blue) is most readily absorbed by the chlorophyll pigments. In contrast, light of 550 nm (green) produced no detectable photosynthesis, suggesting that light of this wavelength is reflected rather than absorbed by the chlorophyll.

Note also that the sample Abstract is informative. The author does not simply say that "Oxygen consumption varied with wavelength.

These results are discussed in terms of the wavelengths that chlorophyll absorbs and reflects." Rather, the author provides a specific summary of the results and what they mean. Be sure that your Abstract is equally informative. Clearly, this section of your report will be easier to write if you save it for last.

PREPARING AN ACKNOWLEDGMENTS SECTION

Most biologists are aided by colleagues in various aspects of their research, and it is customary to thank those helpful people in an Acknowledgments section, the penultimate section of the report. Here is an example that might be found in a typical student report:

> I am happy to thank Casey Deiderich and Steve Untersee for sharing their data with me, and Professor Collin Johnson for helpful discussions concerning the effects of temperature on metabolic rate. Professor C. Orians made me aware of the crucial Lesser and Schick (1989) reference. Finally, I am also indebted to Jean-François Vilain for lending me his graphics software, and to Regina Campbell for teaching me how to use it.

As in the example above, you must include the last names of the people you are acknowledging and indicate the specific assistance that you received from each person named.

PREPARING THE LITERATURE CITED SECTION

In the Literature Cited section, the final section of your paper, you present the complete citations (in alphabetical order, according to last name of the first author of each paper or book) for all the factual material you refer to in the text of your report. This presentation provides a convenient way for readers to obtain additional information about a particular topic, as well as a means of verifying what you have written as fact. Detailed directions for preparing this section are given in Chapter 5 (pp. 74–79).

PREPARING A PAPER FOR FORMAL PUBLICATION

Papers submitted to an editor for possible publication must conform exactly to the requirements of the specific journal you have targeted. Before beginning a manuscript, you must determine which is the most appropriate journal for your work and then read carefully that journal's Instructions for Authors, typically found at the front or back of each issue or, in some cases, at the front or back of several issues each year. Many journals also make this material available online. **It also helps to study similar papers published in recent issues of the targeted journal.** How are references cited in the text? How are they listed in the Literature Cited section? Does the journal permit (or require) subheadings in the Materials and Methods or Results section? If you fail to follow the relevant instructions, your paper may be returned unreviewed; at the least, you will annoy the editor and reviewers.

Do not incorporate figures and tables into the text of the manuscript unless you are required to do so (as for some online journals). Instead, put tables (in numerical order and include the table legends) after the Literature Cited and Acknowledgments sections. Then insert a page of figure captions (in numerical order, with multiple captions per page), and finally, include the figures themselves. These days, manuscripts—including the figures—are usually submitted online; follow the specific instructions given by the targeted journal.

Before submitting your manuscript to the journal's editor, go through your work one last time, and be certain that every reference cited in the text is listed (and correctly so) in the Literature Cited section, and that the Literature Cited section contains no references not actually mentioned in the text.

Your manuscript should be accompanied by a brief cover letter, which should include the paper's title, the names of authors, and short summaries of the research goals and major findings. Here is an example:

Dear Dr. Grassle:

Please consider the enclosed manuscript entitled "Understanding the effects of low salinity on fertilization success and early development in the sand dollar *Echinarachnius parma*," by J.D. Allen and J.A. Pechenik, for publication in *The Biological Bulletin*. Our goal was to determine whether the failure of embryos to develop at low salinities was due to a failure of eggs to be fertilized or to a failure of fertilized eggs to cleave. Our results were surprising, in that reproductive failure at low salinities in *E. parma* is apparently due more to an inability of the fertilized egg to cleave than to an inability of sperm to fertilize eggs. Drs. David

Epel (Hopkins Marine Station, Stanford University) and John Havenhand (Tjärnö Marine Biological Laboratory Sweden) would be especially suitable reviewers for this manuscript. I can send the original figures and photographs immediately upon request.

My contact information follows:

phone (617-627-9999)
fax (617-627-3805)
e-mail: jan.pechenik@tufts.edu

Thank you for your attention.

Note that in the example above, the author recommends specific reviewers for the submitted article. Although editors are happy to have you suggest appropriate reviewers, they won't necessarily take all of your suggestions. Recommend experienced people who will give honest and carefully considered reviews—if the manuscript has problems, you want them found out before publication. Once the paper is published, it's out there forever.

A NOTE ABOUT CO-AUTHORSHIP

Most research papers have multiple authors, all of whom are expected to have made intellectual contributions to the work in addition to any role in collecting the actual data. Intellectual contributions can include designing the study and analyzing and interpreting its results. Authors are also expected to have had a role in writing at least parts of the manuscript and to have read and approved the entire manuscript before its submission to the editor. People who have only helped collect the data or who have only provided equipment or space for the research should be mentioned in the Acknowledgments section but should not be listed as authors.

CHECKLIST FOR THE FINAL DRAFT

TITLE

❑ Title gives a specific indication of what the study is about (p. 209)

ABSTRACT

❑ Background stated in 1 or 2 sentences (p. 211)

❑ Clear statement of specific question addressed and of specific hypotheses tested (p. 211)

- ❏ Methods summarized in no more than 3 or 4 sentences (p. 211)
- ❏ Major findings reported in no more than 2 or 3 sentences (p. 211)
- ❏ Concluding sentence relates to statement of specific question addressed (p. 211)
- ❏ Abstract is a single paragraph; if not, can it be rewritten as one paragraph? (p. 211)

INTRODUCTION

- ❏ Clear statement of specific question or issue addressed (p. 204)
- ❏ Logical argument provided as to why the question or issue was addressed (pp. 206–207)
- ❏ Specific hypotheses are indicated, if appropriate, and a rationale for those expectations is provided (pp. 197, 204–205)
- ❏ Every sentence leads to the statement of what was done in this study (pp. 203, 207)
- ❏ All statements of fact or opinion are supported with a reference or example (p. 206)
- ❏ If appropriate, the rationale for choosing the study system or organism is given (p. 209)

MATERIALS AND METHODS

- ❏ Methods are presented in the past tense (p. 189–190)
- ❏ Design of study or experiment is clear and complete (pp. 158–161)
- ❏ Rationale for each step is self-evident or clearly indicated (pp. 161, 164)
- ❏ Each factor mentioned is likely to have influenced the outcome of this study, and all factors likely to have influenced the outcome are mentioned (pp. 159–160, 164)
- ❏ Precision of all measurements is indicated (p. 160)
- ❏ Includes brief description of how data were analyzed (calculations made, statistical tests used) (p. 161)
- ❏ If appropriate, the field site or study organism is described (pp. 162)

RESULTS

- ❏ Text summarizes important findings in the data and does not simply repeat raw data from the graphs or tables (pp. 188–192)
- ❏ Results are presented in the past tense (p. 189–190)

- ❑ Results are presented in active terms whenever possible (e.g., in terms of what organisms or enzymes did) (pp. 100–102)
- ❑ All general statements are supported with reference to data (and by results of statistical analysis when possible) (pp. 188–189)
- ❑ Major results are presented in words, but their implications are not discussed (pp. 156, 164,190)
- ❑ No raw data are presented (p. 166)
- ❑ Figures are referred to as "Figures" and not as graphs, drawings, or photographs (p. 165)
- ❑ The same data are not presented in both tabular and graphical form within the same report (pp. 168, 187)
- ❑ Every table or graph makes an important and unique contribution to the report (pp. 168, 172)
- ❑ Each figure or table has an informative caption or legend, correctly placed (below figure, above table) (pp. 169, 172)
- ❑ Symbols are used consistently in all figures and are chosen to facilitate interpretation when possible (pp. 169–170)
- ❑ Tables and figures are numbered in the order in which they are first referred to in the paper (p. 188)
- ❑ Each figure or table is self-sufficient; readers can tell what question is being asked, what the major aspects of how the question was addressed are, and what the most important results are without reference to the rest of the paper (pp. 37–41, 169–170, 187–188)
- ❑ Numbers of individuals and numbers of replicates are clearly indicated in the graph, table, caption, or legend (pp. 179, 187–188)
- ❑ The meaning of error bars on figures is clearly indicated in the caption (e.g., 1 standard error about the mean) (pp. 178–188, 187)
- ❑ Results of statistical analyses are correctly incorporated into the text or figures (pp. 64–68)

DISCUSSION

- ❑ Data are clearly related to the expectations and hypotheses raised in the Introduction (pp. 196–199)
- ❑ Facts are carefully distinguished from speculation (p. 199)
- ❑ Unusual or unexpected findings are discussed logically, based on biology rather than apology (pp. 197–199, 201–203)
- ❑ All statements of fact or opinion are supported with references to the literature, data, or an example (pp. 197, 202)

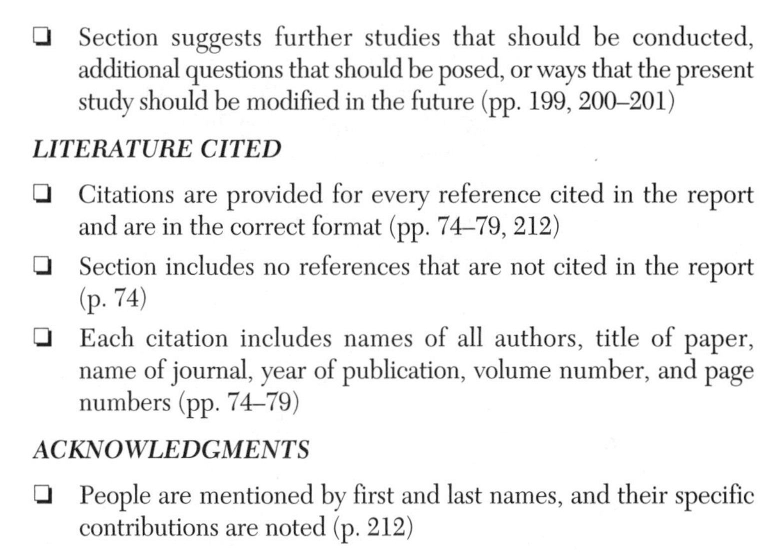

- ❑ Section suggests further studies that should be conducted, additional questions that should be posed, or ways that the present study should be modified in the future (pp. 199, 200–201)

LITERATURE CITED

- ❑ Citations are provided for every reference cited in the report and are in the correct format (pp. 74–79, 212)
- ❑ Section includes no references that are not cited in the report (p. 74)
- ❑ Each citation includes names of all authors, title of paper, name of journal, year of publication, volume number, and page numbers (pp. 74–79)

ACKNOWLEDGMENTS

- ❑ People are mentioned by first and last names, and their specific contributions are noted (p. 212)

GENERAL

- ❑ Text of report is double-spaced (p. 12)
- ❑ First page shows name of author, name of lab section or instructor, and date submitted (p. 12)
- ❑ All information is presented in the appropriate section of the report (pp. 155–156)
- ❑ All pages are numbered (p. 12)

TECHNOLOGY TIP 7

Graphing with Excel

Excel was designed for businesspeople, not biologists, and for graphing it is less flexible and intuitive than some other programs, such as my favorite, GraphPad Prism. However, Excel is so widely used (and misused) by students in biology courses that it is worth pointing out some tips for using it effectively. I assume here that you already have some familiarity with the program.

Entering data. Enter your data in a new spreadsheet. For bar charts, enter the treatment names (the independent variable; e.g., human, dolphin, sea lion, and so on if you were plotting Fig. 24) in the first row. Enter your data starting with the second row.

(*Continued*)

If you are plotting data for a scatter plot (e.g., Figs. 14, 16, and 18), enter data for your x-axis (usually the independent variable, e.g., temperature in Fig. 18) in the first column. In Figure 18, feeding rate data would then be entered in the adjacent column. For Figure 14, you would enter feeding rate data for each of the 3 treatments in adjacent columns. Do not leave space between columns.

Save your worksheet frequently. I recommend including the date on which the study was conducted as part of the title you give the project.

Transforming data. You may wish to transform your data before plotting it. For example, you may need to convert your measurements to different units, or you may wish to plot the logarithms of the data collected. To transform a column of data, first click on an empty column. Click in the first box of the new column and type an equal sign, which tells the program that you are about to enter a formula. Then, select Function within the Insert command on the menu bar. The transform functions you need are found under the Math option. Once you select the desired function, specify the row and values to be transformed. For example, suppose you have 8 values in Column B of your spreadsheet. To take the log (base 10) of those 8 data points and put those new values in Column D, first click in space D1 and type an equal sign. Then, select the Log10 function, enter A1 within the parentheses, and click OK. Next, click on the button in the lower right of the rectangle around space D1, and drag down to space D8. You should see the transformed values in all 8 rows of column D.

Calculating statistics. First, click on an empty rectangle below the column of numbers you wish to work with, and type an equal sign in the box, which instructs the program that you are about to insert a formula. Then, select Function within the Insert command on the menu bar, and choose the correct option (e.g., Average, for calculating the mean, or StDev, for calculating the standard deviation). Next, specify the range of values to use (e.g., B1:B8), and click OK. Once you have calculated your statistic for one column of data, drag the calculation into adjacent columns to calculate the same statistics for those columns.

Making graphs. One of the many endearing idiosyncrasies of Excel is the terminology it uses, which agrees hardly at all with that used by biologists. I present an Excel decoder in Table 8.

Note the tool bar above the worksheet that contains your data. To plot a graph, first highlight the data you wish to plot. Usually

Table 8. The Excel decoder

What It Means	What Excel Calls It
Graph	Chart
x-axis	Category axis
y-axis	Value axis
Area inside the axes of the graph	Plot area
Area outside the axes of the graph but inside the frame	Chart area
Scatter plot (a point graph)	XY (Scatter)
Don't use this!	Line (under Chart Type)
Key (for on-graph explanation of symbols used in graph)	Legend

you will highlight all the numbers entered (by holding the left mouse button and dragging over the screen), but sometimes you will want to select certain columns (e.g., columns of transformed data). Then, alert Chart Wizard (hereafter referred to as CW) of your intentions to plot a graph by clicking on the CW icon (a small but colorful bar chart) toward the right side of the tool bar. When working with the CW, keep the following in mind:

- If you wish to plot a bar graph or histogram, select the Column option. Do not plot 3-dimensional graphs, which can be difficult to read.
- If you wish to produce a scatter plot (e.g., Figs. 14 and 18), choose XY Scatter.
- Do not select Line even when you wish to plot a line graph. Your *x*-axis values will not be properly spaced.
- For scatter plots, you have several suboptions to choose from. Usually you will want to just plot the points without any lines (you can add a regression line later) or to connect the points with straight lines. Occasionally you will want to include smooth curves. **Never choose the options for plotting curves without data points.**
- To add a regression line to a scatter plot, first click on any point in the graph to identify the data set, and then select Add Trendline.

- Use white as a background color for all graphs to maximize contrast and clarity. To do so, double-click anywhere inside the graph, and select the white icon (the default option is gray!). At the same time, select None for Area, and also select None for Border (to remove the frame that Excel otherwise draws around each graph).
- If you will include a key to symbols used in your graph (i.e., if you are displaying more than 1 treatment, requiring the use of at least 2 symbols), use white as the background color for that as well. Double-click anywhere within the key area, and follow the instructions given above.
- If space is available in the figure, move the key into the figure by clicking on it once and then dragging it into the figure.
- In the Chart Options of the Chart menu, leave Chart Title blank. You will enter your title as part of your figure caption later. Be sure to enter the units (in parentheses) after the labels for the x- and y-axes.
- To add error bars to a point or bar, double-click on the point or bar to obtain the Format Data Series menu, and then select the Y Error Bars option. Enter the values of the error bars to be added (e.g., the standard errors, calculated on the spreadsheet using the Function button), in sequence and separated by commas (e.g., 8.4, 12.5, and 5.5), in the Custom rectangle on the menu. Choose the type of display you want, at the top of the menu (usually "plus" for bars, and "both" for points).
- To change the size of the points plotted in a scatter plot, double-click on any point, and change the size in the Patterns part of the menu.
- To modify a graph at a later time, click anywhere inside the graph, and then select the desired option from the Chart item in the menu bar. If you wish to add or remove particular data points, do so first on the spreadsheet, and then use either the Add Data or Source Data options in the Chart menu.
- If you correct a data entry in the spreadsheet, the correction will appear automatically in your graph.

10

WRITING RESEARCH PROPOSALS

What Lies Ahead? In This Chapter, You Will Learn

- How to come up with viable ideas for feasible and convincing research projects
- The importance of anticipating questions and objections that a reviewer might have about your proposed research or how you plan to conduct it
- That a research proposal is a creative undertaking that must be logically developed to be convincing

A research proposal is much like a research or laboratory report, except that it's about work that has not yet been done. In both cases, the Introduction must make a convincing, logical argument—in the case of a report, for why the study was done; in the case of a proposal, for why the study should be done. Similarly, both have a Materials and Methods section; but in one case, you write in the past tense about what you did and why you did it, and in the other, you write in the future tense about what you plan to do and why you plan to do it in a particular way. And both assignments require that you evaluate and synthesize the primary literature—that is, papers presenting original research results rather than articles and books that only summarize and interpret those results. In writing a research proposal, however, you propose to go beyond what you have read; you propose to address a new research question and seek to convince readers that what you propose to do should be done; can be done; and should, in fact, be done exactly how you propose to do it.

Research proposals are excellent vehicles for developing your reasoning and writing skills. This assignment, more than any other, gives you a chance to be creative and to become a genuine participant in the process of biological investigation. Writing a good research proposal is no trivial feat, and the sense of accomplishment you will feel once you are finished is very satisfying.

Research proposals have 2 major parts: a review of the relevant scientific literature and a description of the proposed research. In the first part, **you review the primary literature on a particular topic, but you do so with a particular goal in mind**: You wish to lead your reader to the inescapable conclusion that the question you propose to address follows logically from the research that has gone before. Writing a research proposal rather than a review paper thus helps you avoid falling into the book report trap; once you develop a research question to ask, you should have an easier time focusing your literature review on the development of a single, clearly articulated theme. Developing that theme will take some time and thought, but your writing will then have a clear direction.

In addition to providing you with a convenient vehicle for exploring and digesting the primary scientific literature and for focusing your discussion of that literature, you may find that the research you propose to do can actually be done—and can be done by you. Your proposal could turn out to be the basis for a summer research project, a senior thesis, or even a master's or Ph.D. thesis.

WHAT ARE REVIEWERS LOOKING FOR?

Imagine that you must read 30 to 40 research proposals this weekend and decide which 5 or 6 to fund. Without funding, the proposed work can't be done, so your decisions are important. How would you make them?

Here are some of the features that reviewers (and instructors) look for in evaluating a proposal:

1. Does the applicant know the literature in his or her field and understand it thoroughly?
2. Is the applicant asking good, interesting questions?
3. Do the questions follow logically from what is already known about the topic?
4. Are there specific hypotheses to be tested?
5. Can the hypotheses be tested?
6. Can the hypotheses be tested successfully with the methods proposed?
7. Are potential problems anticipated? If so, does the researcher have plans for dealing with them?
8. If proposing a field study, does the author indicate where the study will be done and why that site was chosen?
9. Does the author indicate sample sizes, numbers of replicates, and how the data will be analyzed?

As are reviewers of job applications (see Chapter 13), reviewers of grant proposals are looking for reasons *not* to fund the proposals they are reading at least as much as they are looking for reasons to fund them. If your writing is not clear, knowledgeable, focused, logical, and careful, why would reviewers think that you will be clear, knowledgeable, focused, logical, and careful in your research?

RESEARCHING YOUR TOPIC

Proceed as you would for researching a review paper (term paper) or essay (see Chapter 7). For this assignment especially, you must have a firm grasp of your subject before plunging into the original, primary scientific literature, so **read the appropriate sections of several relevant textbooks and perhaps a recent review paper before you look elsewhere**. The *Annual Review of...* series is an excellent source of reviews in many fields. The next step should be to browse through recent issues of appropriate scientific journals; your instructor can suggest several that are particularly appropriate to your topic of interest.

Before you roll up your sleeves and prepare to wrestle in earnest with a published scientific paper, read it through once for general orientation. When you begin your second reading of the paper, don't allow yourself to skip over any sentences or paragraphs you don't understand (see Chapter 3, pp. 35–36). Keep an appropriate textbook by your side as you read the primary literature so that you can look up unfamiliar facts and terminology.

I mentioned previously that the results of any study depend largely on the way the study was conducted (pp. 158, 198). We have also seen that although the results of a study are real, the interpretation of those results is always subject to change (p. 164). **The Materials and Methods section and the Results section of research papers must therefore be read with particular care and attention**, as discussed in Chapter 3. Scrutinize every table and graph until you can reach your own tentative conclusions about the results of the study before allowing yourself to be swayed by the author's interpretations. Read with a questioning, critical eye (see Chapter 3).

As you carefully read each paper, pay special attention to the following:

1. What specific question is being asked?
2. How does the design of the study address the question posed?
3. What are the controls for each experiment? Are they appropriate and adequate?
4. How convincing are the results? Are any of the results surprising?

5. What contribution does this study make toward answering the original question?
6. What aspects of the original question remain unanswered?

Reread the paper until you can answer each of these questions. Then ask yourself the following additional question:

7. What might be a next logical question to ask, and how might this question be addressed?

Continue your library research using the references listed at the end of the recent papers you are reading, and perhaps by consulting the ISI Web of Science or one of the other indexing services discussed in Chapter 2. One particularly nice thing about preparing a research proposal is that it's relatively easy to tell when your library work is finished: It's finished when you know what your proposed research question will be and when you know exactly why you are asking that question.

WHAT MAKES A GOOD RESEARCH QUESTION?

You need not propose to cure or prevent any particular disease, rid the world of hunger or parasites, or single-handedly solve any other major humanitarian problem. Rather, **your goal is to pose a specific question that follows in some logical way from what has already been published in your area of interest and that can be addressed by available techniques and approaches** (Figure 31). This can be tricky to accomplish. On the one hand, you might ask a perfectly valid question but

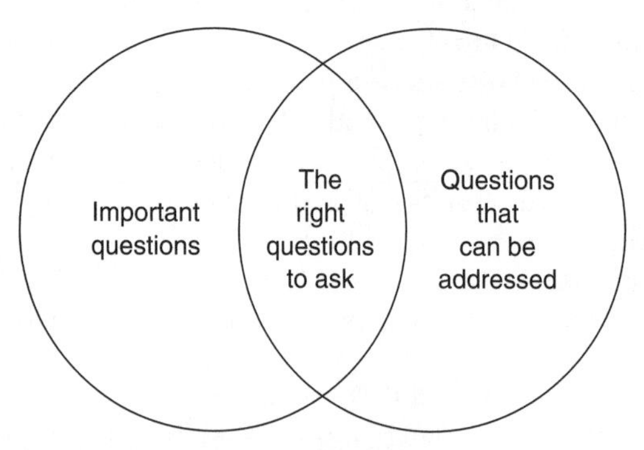

Figure 31. The trick of developing a valid research question. Many questions are easy to answer but are meaningless or too trivial to be worth asking. Many other questions are important but unapproachable by existing methods.

come up with no good way to address it convincingly. It is more common, however, for students to pose addressable questions that are difficult to justify. A question such as "Does music influence plant growth?" can certainly be addressed, but you will have great difficulty convincing readers that the question is worthwhile, because it has no foundation in the scientific literature. What you propose to do must not only be doable; it must also seem like the next most logical question to ask in the area in which you are proposing to work, based on previously published research.

The question you propose must also be within your realm of expertise. You cannot write a convincing proposal on a topic that you do not fully understand.

The trick to asking a good question is to write down lots and lots of questions as you read and as you think about the topic. Many of the questions won't lead anywhere or won't lead anywhere that interests you, but eventually you will come up with something that fits the bill.

WRITING THE PROPOSAL

Divide your paper into 3 main portions: Introduction, Background, and Proposed Research.

Introduction

Give a brief overview of the research being considered, and indicate the nature of the specific question you will pursue, as in the following example*:

> Endurance exercises such as running and swimming can affect the reproductive physiology of women athletes. Female runners (Dale *et al.*, 1979; Wakat *et al.*, 1982), swimmers (Frisch *et al.*, 1981), and ballet dancers (Warren, 1980) menstruate infrequently (i.e., exhibit oligomenorrhea) in comparison with nonathletic women of comparable age, or not at all (amenorrhea). The degree of menstrual abnormality varies directly with the intensity of the exercise. For example, Malina *et al.* (1978) have shown that menstrual irregularity is more common, and more severe, among tennis players than among golfers.
>
> The physiological mechanism through which strenuous activity disrupts the normal menstrual cycle is not yet clear; inadequate fat levels (Frisch *et al.*, 1981), altered hormonal balance (Sutton *et al.*, 1973), and physiological predisposition (Wakat *et al.*, 1982) have each been implicated.

*Modified from a student paper written by A. Lord.

> In the proposed research, I will study 200 female weight lifters in an attempt to determine the relative importance of fat levels, hormone levels, and physiological predisposition in promoting oligomenorrhea and amenorrhea.

Notice that the author of this proposal has not used the Introduction to discuss the question being addressed or to describe how the study will be done. The Introduction provides only (1) general background to help the reader understand why the topic is of interest and (2) a brief but clear statement of the specific research topic that will be addressed.

It helps to write the last sentence of your Introduction first, stating the specific question to be addressed; then write the rest of your Introduction, giving just enough information for the reader to understand why anyone would want to ask such a question. Limit your Introduction to 2 or 3 paragraphs. A detailed discussion of prior research belongs in the Background section of the proposal, and a detailed description of the proposed study belongs in the Proposed Research section of the proposal. Idea mapping (Chapter 6, pp. 84–86) can be an excellent aid to organizing your Introduction.

Notice in the above example that **every factual statement** (e.g., "Female runners ... menstruate infrequently.") **is supported by a reference to one or more papers from the primary literature**. These references enable readers to obtain additional information on particular aspects of the subject and to verify the accuracy of statements made in the proposal (Chapter 5). Backing up statements with references also protects you by documenting the source of information; if the author of your source is mistaken, why should you take the blame? Finally, backing up your statements with references makes you appear well read and knowledgeable.

Background

In this section, you demonstrate your complete mastery of the relevant literature. Discuss this literature in detail, leading up to the specific objective of your proposed research. This section of your proposal follows the format of a good review paper or essay, as already described in Chapter 7 (pp. 132–140). In a proposal, however, **the Background section will end with a brief summary statement of what is now known and what is not yet known** about the research topic under consideration **and will include a clear, specific description of the research question(s) you propose to investigate.**

Here is an example. The author has already spent one paragraph of the Background section describing documented effects of organic pollutants on adults and developmental stages of various marine vertebrates and invertebrates.

EXAMPLE 1

Thus many fish, echinoderm, polychaete, mollusk, and crustacean species are highly sensitive to a variety of fuel oil hydrocarbon pollutants, and the early stages of development are especially susceptible. However, many of these species begin their lives within potentially protective extra-embryonic egg membranes, jelly masses, or egg capsules (Anderson *et al.*, 1977; Eldridge *et al.*, 1977; Kînehcép, 1979). The ability of these structures to protect developing embryos against water-soluble toxic hydrocarbons has apparently never been assessed. The egg capsules of marine snails are particularly complex, both structurally and chemically (Fretter, 1941; Bayne, 1968; Hunt, 1971). Such capsules are typically several mm to several cm in height, and the capsule walls are commonly 50–100 μm* thick (Hancock, 1956; Tamarin and Carriker, 1968). Depending on the species, embryos may spend from several days to many weeks developing within these egg capsules before emerging as free-swimming larvae or crawling juveniles (Thorson, 1946).

Little is known about the tolerance of encapsulated embryos to environmental stress, or about the permeability of the capsule walls to water and solutes. Kînehcép (1982, 1983) found that the egg capsules of several shallow-water marine snails (*Ilyanassa obsoleta*, *Nucella lamellosa*, and *N. lapillus*) are permeable to both salts and water, but they are far less permeable to the small organic molecule glucose. Capsules of at least these species are thus likely to protect embryos from exposure to many fuel oil components.

In the proposed study, I will (1) document the tolerance of early embryos of *N. lamellosa* and *N. lapillus*, both within capsules and removed from capsules, to the water-soluble fraction of Number 2 fuel oil; (2) determine the general permeability characteristics of the capsules of these 2 gastropod species to see which classes of toxic substances might be unable to penetrate the capsule wall; and (3) use radioisotopes to directly measure the permeability of the capsules to several major components of fuel oil.

Notice that by the end of the first paragraph in this example, the author already gives us a very clear indication of what the proposal will be about. This section of your proposal has the potential to lead a double life: It can later serve as the basis for the Introduction and Discussion sections of a thesis or research article. In fact, **the Introduction section of any well-written, published research paper can serve as a model for what you are trying to accomplish in the Introduction section**

*μm = micrometers (10^{-6} meters).

of your proposal. Only the tenses will differ: This section of the published research paper will be written in the past tense, ending with a concise summary of what the researchers set out to do (Chapter 9), while your Introduction will end instead in the future tense, with a concise summary of what you *plan* to explore.

Proposed Research

This portion of your proposal has 2 interrelated parts: (1) what specific question(s) will you ask? and (2) how will you address each of these questions? Different instructors will put different amounts of stress on these 2 parts. For some of us, the formulation of a valid and logically developed question is the major purpose of the assignment, and a highly detailed description of the methods will not be required. For such an instructor, you may, for example, propose to extract and separate proteins without actually having to know in detail how this is accomplished. But other instructors may believe that your mastery or knowledge of methodological detail is as important as the validity of the questions posed. Both approaches are defensible, depending largely on the nature of the field of inquiry, on the level of the course being taken, and on the amount of laboratory experience you have had. Be sure you understand what your instructor expects of you before preparing this section of your paper.

Before you begin to write this section of your proposal, I strongly recommend that you sketch a flowchart of your proposed study, as shown in Figure 11 (p. 152). This will help you organize your thinking and will also serve as a template for your writing. Often it is helpful to include such a flowchart in your proposal, making it easy both for the reader to grasp the complete experimental design and for you to write about it.

As you describe each component of your proposed research, **indicate clearly what specific question each experiment is designed to address**, as in the following 2 examples:

> To see if there is a seasonal difference in the amount of hormone present in the bag cells that induce egg-laying in *Aplysia californica*, bag cells will be dissected out of mature individuals each month and...

> Before the influence of light intensity on the rate of photosynthesis can be documented, populations of the test species (wild columbine, *Aquilegia canadensis*) must be established in the laboratory. This will be done by ...

If the proposed research has several distinct components, it is helpful to separate them using subheadings. Your first subheading might read, for example, "Collecting and maintaining *Aplysia californica* adults," while a second subheading might read, "Isolating and homogenizing bag cells," and a third might read, "Assaying for hormonal activity" (see Chapter 9, pp. 162–163, for additional examples).

Model your Proposed Research section on the Materials and Methods section of any well-written, published research article. Again, only the tenses will differ. Consider this recent paragraph from the Materials and Methods section of a paper by Barresi *et al.* (2000)* and its modification into a format suitable for a Proposed Research section. I have bold-faced the words that differ in the two versions.

> We **created** genetic mosaics between wild-type and mutant embryos essentially as described (Ho and Kane, 1990). Donor embryos **were injected** at the 1- to 4-cell stage with lysinated rhodamine dextran (10,000 kDa, Molecular Probes). Between 3 h and 5 h, 10–50 cells **were transplanted** from these embryos into similarly staged embryos. Transplant pipettes **were made** on a sanding disk constructed from a discarded hard drive coated with diamond lapping film. Transplantations **were done** using an Olympus SZX12 dissecting microscope. At 24 h, the $smu^{-/-}$ embryos **were identified** on the basis of partial cyclopia and the U-shape of their somites. Embryos **were fixed** and sectioned on a cryostat; sections **were then labeled** with F59 to identify muscle fiber type (Devoto *et al.*, 1996). Slow and fast muscle fibers derived from donor cells **were counted** in every third section in all cases.

Here is the same material rewritten as it would appear in a proposal:

> We **will create** genetic mosaics between wild-type and mutant embryos essentially as described (Ho and Kane, 1990). Donor embryos **will be** injected at the 1- to 4-cell stage with lysinated rhodamine dextran (10,000 kDa, Molecular Probes). Between 3 h and 5 h, 10–50 cells **will be** transplanted from these embryos into similarly staged embryos. Transplant pipettes **will be** made on a sanding disk constructed from a discarded hard drive coated with diamond lapping film. Cells **will be** transplanted using an Olympus SZX12 dissecting microscope. At 24 h, the *smu*$^{-/-}$ embryos **will be** identified on the basis of partial cyclopia and the U-shape of their somites. Embryos **will be** fixed and sectioned on a cryostat; sections **will then be** labeled with F59 to identify muscle fiber type (Devoto *et al.*, 1996). Slow and fast muscle fibers derived from donor cells **will be** counted in every third section in all cases.

---

*From Barresi, M.J.F., Stickney, H.L., Devoto, S.H. 2000. The zebrafish *slow-muscle-omitted* gene product is required for Hedgehog signal transduction and the development of slow muscle identity. *Development* 127: 2189–2199.

Here's a happy thought: If you end up performing the research proposed, you will already have your Materials and Methods section nearly ready for publication!

Citing References and Preparing the Literature Cited Section

Cite references directly in the text by author and year, as in the examples given earlier in this chapter (see also pp. 69–74). The Literature Cited section of your proposal is prepared as described in Chapter 5 (pp. 75–79).

TIGHTENING THE LOGIC

Read your proposal aloud, slowly and thoughtfully, before deciding that your work is finished. If you listen as you read, you can often catch logical and typographical errors that you might otherwise miss.

In rereading your description of what you propose to do and how you propose to do it, **try to envision the specific objections that an interested but critical reviewer would raise**. Can you argue those objections away? Do so in your presentation, if possible. If not, can you modify your approach or add additional components to the study that will address those specific objections? Perhaps you need to add additional controls or additional experiments, or to modify your experimental design. In some cases, you may need to modify the question that you are proposing to address.

It is the reviewer's job—and your instructor's—to find flaws in your proposal. Try to be your own harshest critic, and find and fix as many as you can before anyone else judges your work.

THE LIFE OF A REAL RESEARCH PROPOSAL

This is no idle exercise; the formal proposals written by practicing biologists are prepared pretty much as described, except that each proposal must adhere strictly to the particular format (major headings, page length, font size, width of margins) requested by the National Science Foundation, National Institutes of Health, or other targeted funding agency. Proposals must be submitted by specified due dates or they will not be considered—no excuses are accepted. Copies of the proposal are then sent out to perhaps 6 to 10 other biologists for anonymous reviews. A panel of still other biologists then meets to discuss the proposal and the reviews, and to then make its own evaluation. If your case is well argued, you may receive funding; if it is not well argued, there is little hope. Learning to write a tightly organized and convincing proposal now will surely make your life easier later, no matter what career ultimately attracts you: Sooner or later you will probably need to convince someone of something, in writing (see, e.g., Chapter 12).

CHECKLIST

(see also the checklist at the end of Chapter 6, on revising)

- ❑ Title gives specific indication of the proposed work (pp. 140–141, 209–211)
- ❑ Introductory material leads to a clear statement of the specific goal(s) and hypotheses (pp. 225–227)
- ❑ The questions posed follow logically from previous work in the area of interest (pp. 222, 224–225)
- ❑ The logic behind all hypotheses presented is made clear (pp. 197, 205, 222–223, 230)
- ❑ Final paragraphs of the Introduction and Background sections address the issues raised in the introductory paragraphs (pp. 226–227)
- ❑ All statements are supported by reference, data, or example (pp. 7, 71–73, 196–197, 226–228)
- ❑ Proposed methods will address the questions posed and are designed to distinguish among all alternative hypotheses (p. 222)
- ❑ A rationale is provided for each step proposed (p. 228)
- ❑ Controls are appropriate and clearly indicated
- ❑ Sample sizes and number of replicates per treatment are indicated
- ❑ Plans for data analysis are clear (pp. 163–164, 222)
- ❑ Each sentence follows from the preceding sentence and leads logically to the one that follows (pp. 102–106)
- ❑ Work has been carefully proofread and revised according to the guidelines presented in Chapter 6 (pp. 80–114)
- ❑ Citations are provided for every reference cited in the report (pp. 74–75)
- ❑ Each listing in the Literature Cited section includes names of all authors, title of paper, year of publication, volume number, and page numbers and is in the correct format (p. 74–79)
- ❑ Text of the report is double-spaced (p. 12)
- ❑ All pages are numbered (p. 12)

11

PRESENTING RESEARCH FINDINGS: PREPARING TALKS AND POSTER PRESENTATIONS

Biologists often attend scientific meetings to share their research progress with others in related fields. Presentations at these conferences most commonly take the form of either oral presentations or poster presentations. You may also be required to present information orally or through posters in your courses.

ORAL PRESENTATIONS

Like written papers, oral presentations must be clearly organized. Indeed, **in writing any paper**—summary, critique, research report, research proposal, or literature review—**it typically helps to think first in terms of giving a clear talk**. An oral presentation differs from a written presentation in one important respect: A printed page can be read slowly, pondered, and reread as often as necessary, until all points are understood, while an oral report offers the listener only one chance to grasp the material. An analogy can be made with music. Before about 1910, music had to be liked at the first hearing; composers knew that if their audience was not captivated during the first performance, that performance might well be the last. Only when the phonograph was invented could a composer sustain a career by intentionally delivering music intended to grow on its audience.

Since **an oral presentation goes past the listener only once**, it must be well organized; logically developed; stripped of details that divert the listener's attention from the essential points of the presentation; and delivered clearly, smoothly, and with enthusiasm.

Talking about Published Research Papers

As a class assignment, you may be asked to talk about a published research paper, perhaps one that you are reading for a larger writing assignment due later in the semester. Just like written work, a talk can be effective only if you fully understand your topic. You'll want to skim the paper that you are presenting once or twice for general orientation; consult appropriate textbooks for background information as necessary; and pay particular attention to the Materials and Methods section and to the tables, graphs, drawings, and photographs in the Results section. When you can summarize the essence of the paper in one or two sentences (see Ch. 3, pp. 39–43), you are ready to prepare your talk.

Writing the Talk

There are many ways to give a good talk: There is no single template that you can mindlessly "plug into." But your goal in preparing a talk is identical to your goal in a written assignment: You seek to capture the essence of the research project—why it was undertaken, how it was undertaken, and what was learned—and to communicate that essence clearly, convincingly, and succinctly. Keep these goals in mind at all times, along with these guidelines:

1. **Do not simply paraphrase** the Introduction, Materials and Methods, Results, and Discussion sections of the paper that you are presenting. If you want to keep your audience awake, you will have to reorganize the paper's content to make the presentation clear and interesting.
2. **Introduce your talk** by providing background information, drawing from the Introduction and Discussion sections of the paper that you read and from outside sources if necessary, so that your listeners can appreciate why the study was undertaken. You want to convince listeners that what you are talking about is worth listening to. **End your introduction with a concise statement of the specific question** or questions addressed in the paper.
3. **Focus the rest of your talk on the questions addressed and the results achieved.** If you use note cards to help you focus on the most important questions and results, don't forget to number the cards, so that they can be easily reassembled if you drop them.

4. **Draw conclusions as you present each component of the study** so that you lead your audience logically from one part of the study to the next. Integrate the Materials and Methods and Results sections to form a continuous story. If you are discussing several experiments from a single paper, state the first specific question, briefly describe how it was addressed, and present the key results; then, lead into the second specific question, describe how that question was addressed, present the key results, lead into the next question, and so forth. For example:

    The oyster larvae grew 20 μm/day when fed diet *A*, 25 μm/day when fed diet *B*, and 65 μm/day when fed a combination of diets *A* and *B*. This suggests that important nutrients missing in each individual diet were provided when the diets were used in combination. To determine what these missing nutrients might be ...

    Lead your listeners by the nose, from point to point.
5. **Be selective.** Much of what is appropriate in a research paper is not appropriate for a talk. Since listeners have only one chance to get the point, some of the details in the paper—particularly methodological details—must be pruned out in preparing the oral presentation. **Include only the details needed to understand what comes later**. If, for example, you will not discuss the influence of animal or plant age on the results obtained, don't include this information in your talk. Similarly, if later in your talk, you will not discuss that samples were mixed on a shaker table, then omit it at the outset. If such tidbits are important to some listeners, they will probably emerge during the question period.
6. **Include visual aids.** PowerPoint and similar software enable you to create summary tables, maps, diagrams, graphs, and flowcharts with relative ease. **A diagram or flowchart of the experimental protocol**, along the lines of that shown in Figure 11 (p. 152), **can help listeners follow the plan of a study.** Try to summarize data in a few simple graphs, even when those data were presented in the original paper as complicated tables. **Use large fonts** for everything, including legends, axis labels, and text on graphs. Focus your presentation on the major trends in the data, and omit anything that fails to help you make your point clearly.
7. **Conclude your talk by summarizing the major points**. If appropriate, include a summary slide of major results. You may

wish to end your talk by suggesting a research question or two that should be asked next. Do not set out to discredit the authors in your summation. Rather, end on a positive note, reinforcing what you want your audience to remember.

8. **Be prepared for questions about methods.** Listeners often ask about interpretations of the data; to answer these questions, you must be thoroughly familiar with how the study was conducted.

Giving the Talk

Knowing what you are going to say and how you are going to say it will almost always lead to a successful talk. Hesitation, vagueness, and searching for words all suggest a lack of understanding and risk losing the attention of your audience. **Practice delivering your talk**—either from a written text or from detailed notes—until you can speak smoothly while maintaining eye contact with your listeners.

Take your time. Especially with PowerPoint presentations, reveal key aspects of a single figure slide by slide, as you talk about each one, rather than showing completed graphs or tables. For each data slide, always take the time to **orient viewers to the axis labels or column headings before plunging into the results**. You might say, for example, "Here we see adenylate cyclase activity on the *y*-axis, in picomoles of cyclic AMP produced per minute per milligram of heart tissue, as a function of time after adding the peptides, up to 1.5 hours." Remember, your audience has not seen these displays before; if you don't first explain what they are looking at, you may be blithely talking about how interesting the results are while your listeners are still figuring out the axis labels. Point to exactly what your listeners should be looking at as you talk.

Make the data work for you. Draw listeners' attention to the specific aspects of graphs or tables that present the points you wish to make. Don't simply say, "This is clearly shown in the graph." Rather, say, "For example, you can see that while all the animals that were fed on diets *A* and *B* grew at comparable rates, those fed on diet *C* . . . " Also, point to the data as you speak, but **be careful when using laser pointers.** These pointers often show up poorly on screens, especially against certain background colors. Moreover, any nervousness you are feeling will translate into a very conspicuous tremor on the screen. If you do use a laser pointer, point it first, turn it on, point out the relevant feature, and then turn it off. **Don't distract your audience by waving an active laser pointer around while you talk.**

Explain unfamiliar terms, and write them down. Use a PowerPoint slide to define unfamiliar terms. Avoid acronyms as much as possible, especially if a term is used only once or twice in your talk (refer to *neural cell adhesion molecules,* not to *NCAMs).* Your goal, after all, is to communicate.

Dos and Don'ts for Oral Presentations

- **DO sound interested in what you are saying**. No matter how many times you have practiced or how familiar you are with this material, remember that your audience has never heard it before. If you seem bored with your presentation, you will almost certainly bore your audience. Ask yourself what you find especially interesting about the material that you are presenting, and then try to make that material interesting to the listeners as well.
- **DON'T put the complete text of your talk on your slides.** You want the audience to be *listening* to you, not reading your notes.
- **DON'T mumble.** Make eye contact with your listeners. Be careful not to talk to your PowerPoint slides or the blackboard.
- **DON'T automatically refer to the author of a paper as *he*.** Many papers are written by women, and many are written by two or more researchers.
- **DO end your talk gracefully, not abruptly.** Let your audience know when you are nearing the end by saying something like, "Finally . . . " or "In summary . . . " Bring your talk to an end with a clear, concise summary. You may want to acknowledge people who gave you advice, let you use their equipment and supplies, or helped in other ways, and then say something like, "Thank you. I would be happy to answer any questions."
- **DON'T exceed the time allotted.** You will lose considerable goodwill by rambling on beyond your time limit. Practice your talk in advance to make sure it will fit within the allotted time. It's often helpful to indicate in your notes about where you should be after half the allotted amount of time has gone by; then you can speed up a little, or slow down a little, for the rest of your presentation as necessary.
- **DO paraphrase each question before answering it,** to make sure the whole audience hears and understands the question, and to buy yourself a few precious seconds to think. Then, address your answer to the entire audience, not just to the person asking the question.

- **DON'T feel compelled to answer questions that you don't understand.** Politely ask for clarification until you figure out what is being asked—and don't be afraid to admit that you don't know the answer. You can easily work your neck into a noose by pretending to know more than you really do; nobody expects you to be the world's authority on the topic you are presenting. Simply saying, "I don't know" is the safest way to go.

Talking about Original Research

When you are talking about your own research, the guidelines for an oral presentation are essentially the same:

- Present the background information that listeners will need to understand why you addressed the particular question or issue you chose.
- Clearly state the specific question or issue that was considered.
- **Focus on the results of previous studies** when presenting the background information **and on your own results** when giving the rest of the talk.
- Draw your conclusions point by point as you discuss each facet of the study, showing how each observation or experiment led to the next aspect of the work.
- End the talk by summarizing your major findings with their potential significance, and perhaps with a brief suggestion of what you might do next to further explore the issue you raised at the start of your talk.

Talking about Proposed Research

The difference in talking about proposed rather than published or original research is that you will have more literature to review. Otherwise, the preparation and delivery of your talk should follow all of the points detailed earlier in this chapter. Highlight a few key papers that clearly show why the question you wish to address is a worthwhile and logical one, and focus on the results of the studies you discuss. Then, state the specific question you plan to address in your own work, being sure that this question follows logically from the work you have just summarized. Finally, describe the approach you will take, focus on what you will do, and make clear what each step of the study is designed to accomplish. Conclude by briefly summarizing how the proposed work will address the question under consideration.

The Listener's Responsibility

Few things in life are as disappointing as putting your heart and soul into preparing and then delivering a talk to an apparently indifferent audience. **When you are a member of that audience, you bear a responsibility to listen closely.** Try to formulate at least one question about something you didn't understand, something you thought was particularly interesting ("I was amazed by the ability of those polar fishes to keep from freezing. Do local fish produce the same kind of biological antifreezes in the winter?"), or something you found unusual in the data ("In that table you showed us, why were the arsenic concentrations so high in the control animals' tissues?").

Even if the speaker can't answer your question, he or she will detect your interest in the talk and feel flattered that you cared enough to have paid so much attention. And your question may well stimulate interesting questions from other audience members.

Preparing Effective Visuals

Talks are almost always improved by the use of clear, effective visuals. A figure or table that works fine in a published paper, however, may not work nearly as well as a visual aid for a talk. When reading a journal article, readers can scrutinize the data for as long as required, even over several cups of coffee if necessary. But for a talk, the audience needs to understand the visual quickly, so that you can concentrate on the results. If a figure is too complicated or too difficult to read, you may finish talking about the results while your listeners are still trying to figure out what your axes are! **Make your visual aids simple and clear.** They should ease communication, not hinder it.

Figure 32, for example, was developed for publication. It concerns seasonal variation in initial carbon content and juvenile growth rate of an intertidal barnacle, *Semibalanus balanoides*. Combined with its caption, the figure is a perfectly acceptable, self-sufficient summary of the data. However, it would not work well for a talk. For one thing, people sitting beyond the front row would probably not be able to read the axis labels, let alone the caption. And you don't want them reading the caption while you're talking anyway: You want them listening to what you are saying.

Figure 33 shows how this graphic might be modified for use in an oral presentation. By using 2 lines instead of 1 for the dates (x-axis) and the y-axis titles, and by shifting the y-axis titles to separate sides of the graph, we can use a much larger, more readable typeface. I have also identified

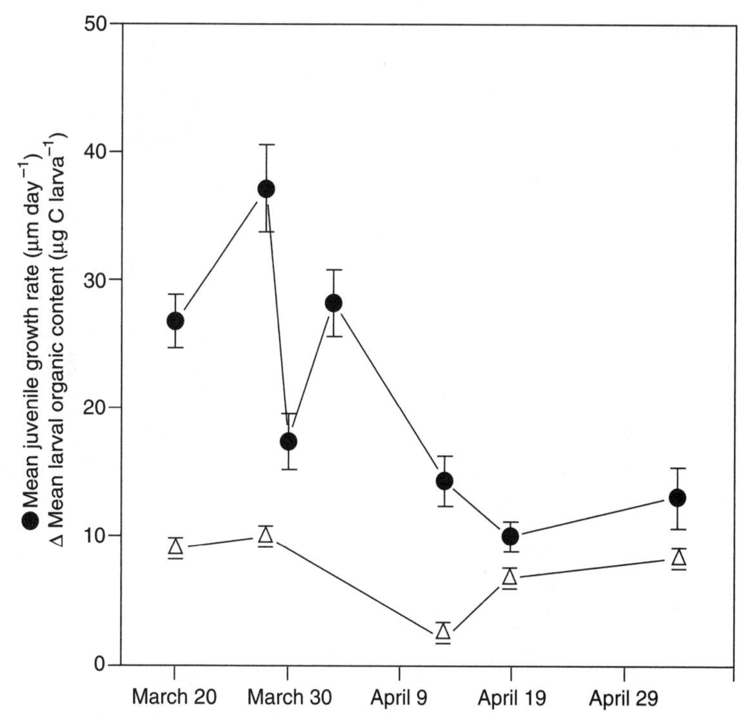

Figure 32 A graph with its figure caption, designed for publication in a formal research paper.
Courtesy of J. Jarrett.

the 2 curves directly on the graph, rather than on the left and have added the name of the species along with where and when the data were collected. Finally, I have removed the figure caption. Listeners—even those at the back of the room—should now have little difficulty reading and understanding the illustrated data. **In designing your visual aids, always consider the person sitting at the back of the room.**

Using PowerPoint

Tools available through programs like PowerPoint allow us to create an astonishing array of visuals—spectacularly colorful, with palm trees and other embedded images and photographs in the background, and even animated text. You can show photographs of your study site and study organisms taken with a digital camera or obtained from Web sites or scanned from books or

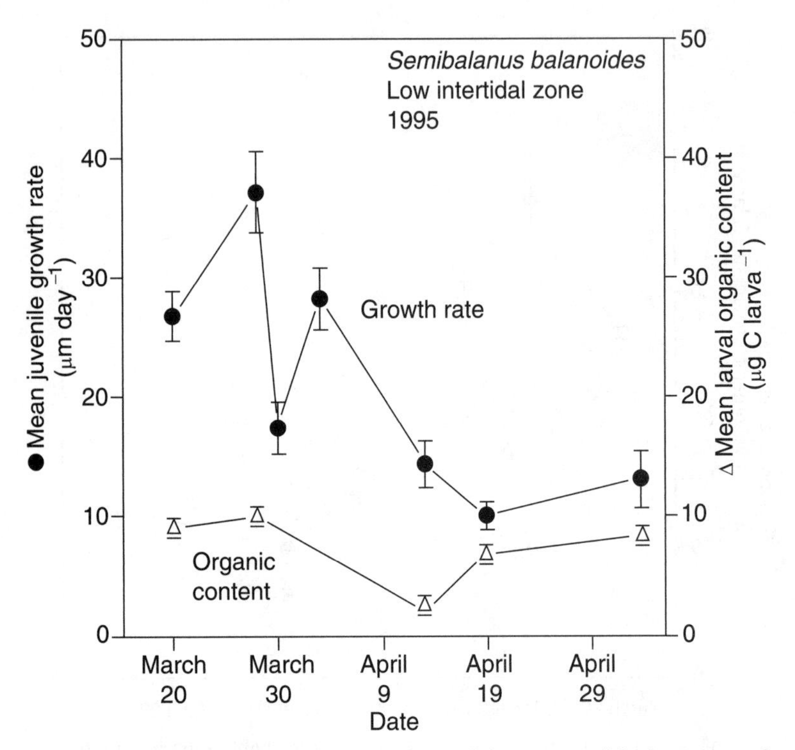

Figure 33 A revised version of Figure 32, designed for use as a slide during an oral presentation. Note that some of the information contained in the figure caption of Figure 32 is now placed directly on the graph.

Courtesy of J. Jarrett.

magazines, and you can add labels and pointers to highlight particular features as you talk. For example, in preparing Figure 33, you might initially display only the left axis and its associated data and then, after talking about those data, add the right axis and its associated data as you begin to talk about them. You can even show movies and animations, bringing your organisms to life for the audience and showing, for example, aspects of the specific behaviors that you studied.

To the extent that you are using PowerPoint® in ways that increase your ability to communicate your research, it's great. (See Appendix C for Web sites offering additional help with creating effective PowerPoint® presentations.) However, **you must not lose sight of your overriding goal, which is to communicate information. So be careful: The technology is impressive, and seductive, but it can work both for and against you.**

Common PowerPoint Errors

Computerized presentations can be disappointing, and even downright disastrous, for several common reasons:

1. **Overreliance on the available resources.** All that color, all those images, all that animation! It's easy to be seduced by all the possibilities. But animated numbers, amusing clip art, and other irrelevant images can be distracting to listeners or, even worse, can suggest that you aren't taking your talk seriously. And blue print on a purple background or black print against a red background may look striking on your computer screen, but it will be almost (or totally) unreadable in presentation. **Keep in mind that maximum clarity is achieved with maximum contrast**, and the best contrast is achieved by putting black letters on a white background (or yellow letters on a dark blue background). Remember, too, that many people are color-blind and will not be able to distinguish among some colors.
2. **Too much text.** Too many speakers simply transcribe their lecture notes onto their PowerPoint slides, crowding each slide with information. But if you give me all that material to read, I'll read it—or try to. I certainly won't be listening to you. If you're saying exactly what I'm reading, then I suppose it doesn't really matter that I've tuned you out... but then again, doesn't that make you superfluous? A good general rule of thumb is 3 bullets to a slide, and 3–4 words to a bullet.
3. **PowerPoint as teleprompter.** If you are constantly reading from your slides to prompt your talk, you'll necessarily have your back to your audience. There is no quicker way to cause an audience to lose interest in what you're saying. Talk to your audience. Make eye contact.
4. **Letting style substitute for substance.** If, when preparing your talk, you are focused mostly on the software and its various bells and whistles, you may not be paying enough attention to analyzing your data or double-checking your analyses; to thinking about what you did, why and how you did it, what the most important results are, and what they mean; or to reading more background information and more research papers, or reading them more carefully. It's fine to be high tech and colorful, but be sure you also have something to say: Style is no substitute for substance. In one of the most compelling talks I've ever heard, the speaker spent 50 minutes talking about different aspects of a single slide that he projected at the start of his talk. It was masterful. **Make the substance of your talk the star of your show.**

CHECKLIST FOR BEING JUDGED

Many scientific societies give awards for the best student presentations at annual meetings. Knowing the criteria used by the judges can help you prepare a more effective presentation. Your instructor will probably be looking at the same sorts of issues in judging an in-class presentation:

- ❑ Was a specific research question clearly stated?
- ❑ Was all background information relevant? Was it sufficient to understand why the question was posed? Did it lead logically to the question stated?
- ❑ Was the talk free of unnecessary or unexplained jargon?
- ❑ Did the speaker deal thoroughly with one issue at a time?
- ❑ Did the methods described follow logically from the questions posed?
- ❑ Were methods presented in the right amount of detail, including sample sizes, numbers of replicates, and use of controls? Was there too little detail? Too much detail? Was the design of the study easy to follow?
- ❑ Were procedures, concepts, and methods of data analysis presented clearly?
- ❑ Were graphs and tables easy to read, uncluttered, and easy to follow?
- ❑ Did the speaker take sufficient time to orient listeners to the data before describing the results?
- ❑ Did the speaker lead listeners through the results and to appropriate conclusions?
- ❑ Was the talk delivered clearly, at an appropriate pace, without reading word-for-word from notes, and without distracting mannerisms?
- ❑ Had the speaker clearly practiced the talk in advance?
- ❑ Did the speaker's talk fit the allotted time?
- ❑ Did the speaker maintain eye contact with audience members?
- ❑ Did the speaker respond well to questions from the audience?

POSTER PRESENTATIONS

For many years, the standard meeting format was a series of 10- to 15-minute oral presentations followed by 5 minutes for questions. However, as the number of meeting participants has increased dramatically without any increase in meeting length, oral presentations are giving way to *poster presentations*, in which displays containing both text and data are lined up in rows, like billboards, for all to see. Each display represents the research of one person or team. Each poster is usually displayed for only half a day, and 50 or more posters—sometimes even many hundreds—may be on display at any one time, **each competing for the attention of meeting attendees**.

The advantage of poster presentations is that many biologists can "talk" about their research simultaneously in a single room, while "listeners" can have detailed conversations with the authors of the posters they find especially interesting. The disadvantage of poster presentations is that the "speaker" no longer has a captive audience: Poster sessions are like flea markets, complete with all the noise and crowds. **To be successful in "selling" your information, you must plan carefully to create a display that captures the attention of browsers** and then leads them through an especially clear, logical, and interesting presentation of the research and its major findings. Otherwise, much of your potential audience will simply pass you by, lured elsewhere by another person's more compelling presentation.

How do you create a poster that will make people stop and read, and from which even the casual reader can take away something of substance? **Plan a two-pronged attack:**

1. Limit the amount of information you present.
2. Arrange the information advantageously.

All too often, posters display what is essentially a full scientific manuscript—complete with formal Introduction, Materials and Methods, Results, and Discussion sections—enlarged and hung up for view, page by page. It is simply not reasonable, however, to expect people to read through dozens of research papers during the hour or so they may spend at a particular poster session. When creating a poster presentation, keep the following guidelines in mind:

- **An effective poster presentation includes less detail than a formal publication or even a talk.** Your poster should be designed to inform people both within and outside your immediate field about what you have done and what you have found, and it should provide a basis for discussion with those who wish to find out more about your work.
- **An effective poster presentation highlights the major questions asked, the major results obtained, and the major conclusions drawn,** and it does so using the least possible amount of text.

Normally, a poster (or an oral presentation) would precede publication, but we can learn something of value by comparing a published work with its corresponding poster. Let's create a poster presentation based on a paper published by Richard K. Zimmer-Faust and Mario N. Tamburri in the journal *Limnology and Oceanography* (vol. 39: 1075–1087). This paper reports a series of experiments defining the chemical cue that causes the swimming, microscopic larvae of oysters to stop swimming and settle to the bottom of the sea in preparation for metamorphosing to the more familiar immobile (and eventually highly edible) juvenile stage.

You attract an audience (or not) with your title, which should appear in large letters at the top of your poster, and **which indicates clearly both the question addressed in the study and the key finding of the study.** The title of the published paper, "Chemical identity and ecological implications of a waterborne, larval settlement cue," is too general to be compelling as a poster title; the passerby who reads only the title leaves with nothing of substance. The poster will attract more attention with a more revealing title: "Oyster larvae settle in response to arginine-containing peptides."

The rest of the poster should focus on the most important results. The published paper contained 7 figures and 4 tables; our poster will display only 3 of the figures and none of the tables. To make it as easy as possible for viewers to extract the essential information we wish to convey, **each of these figures should be self-sufficient**. Each should have clearly labeled axes, contain definitions of any symbols used, and be accompanied by clear indications of the specific question being addressed and the major results found.

The bigger difficulty in achieving a completely self-sufficient figure is in explaining how the experiment was performed or how the observations were made. **Do not include a detailed, formal Materials and Methods section.** Instead, for each figure either (1) list the major steps taken, in numerical order or (2) present a flowchart summary of the steps taken.

You may wish to include a more detailed description of the methods on a one-page handout that particularly interested biologists may take with them, but the poster itself should not be cluttered with such detail. If you do create handouts, be sure they include your name, school affiliation and mailing address, email address, and the poster title.

Layout of the Poster

There are a number of ways to successfully present your research, and different sorts of research may require somewhat different presentation strategies. My goal here is to demonstrate key principles. If all goes well, you should be able to see examples of posters that illustrate many of the suggestions given here on my Web site: http://ase.tufts.edu/biology/faculty/pechenik/ (click on Writing about Biology).

Notice that our poster (Fig. 34) is divided into three major sections, each highlighting one key issue, and that **each section is separated from the other sections by substantial space**. Readers can easily follow the logic of the presentation by scanning from left to right, section by section—there is never any question of where to look next. Even casual readers will find it hard to miss the point.

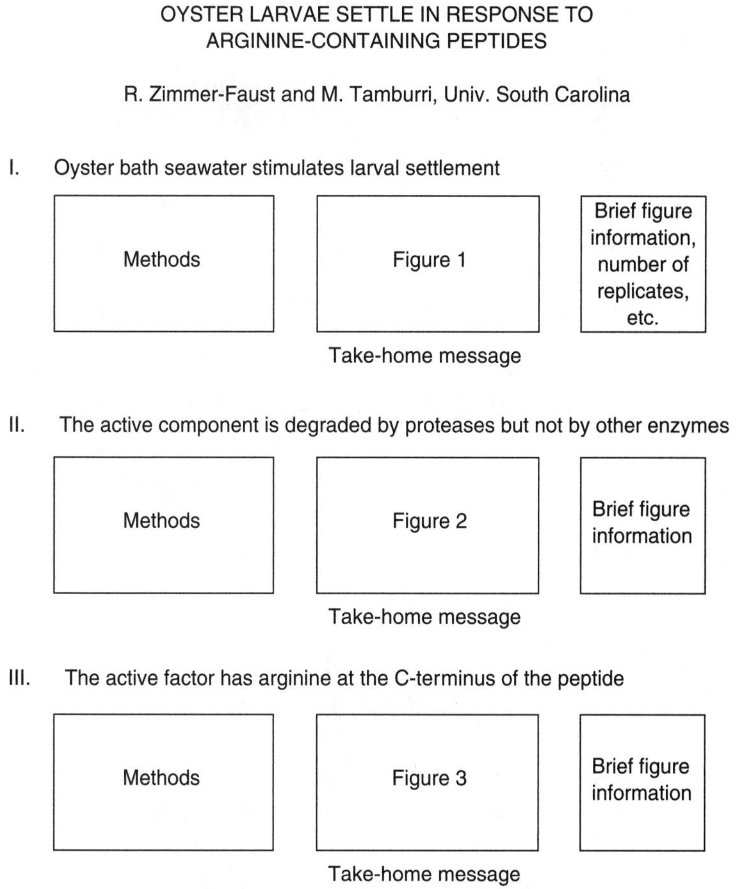

OYSTER LARVAE SETTLE IN RESPONSE TO
ARGININE-CONTAINING PEPTIDES

R. Zimmer-Faust and M. Tamburri, Univ. South Carolina

I. Oyster bath seawater stimulates larval settlement

Methods	Figure 1	Brief figure information, number of replicates, etc.

Take-home message

II. The active component is degraded by proteases but not by other enzymes

Methods	Figure 2	Brief figure information

Take-home message

III. The active factor has arginine at the C-terminus of the peptide

Methods	Figure 3	Brief figure information

Take-home message

Figure 34 General layout of the poster. The goal is to make it easy for readers to see what was done and what was discovered.

Let's fill in the entire top section of the poster, entitled "Oyster bath seawater stimulates larval settlement." Our Methods section might look like this:

Methods:

1. Incubate 8 adult oysters (*Crassostrea virginica*) in 16 liters of artificial seawater for 2 hours.
2. Adjust pH of oyster-conditioned seawater and control seawater to 8.0; adjust salinity to 25 psu.
3. Separate oyster-conditioned seawater and control seawater samples into 3 molecular size-fractions by dialysis.
4. Expose oyster larvae to both solutions.
5. Videotape larval behavior; determine number of individuals settling on bottom of containers by end of 3 minutes.

Here is the same Methods information presented in flowchart format:

<u>Methods</u>

Incubate 8 adult oysters in 16 L of artificial seawater

↓ 2 hours

Adjust pH to 8.0, salinity to 25 psu

↓

Distribute conditioned water into a glass dish

↓

Distribute control seawater into another glass dish

↓

Add sixty 20-day-old oyster larvae to each dish

↓

Videotape for 3 minutes

↓

Assess numbers of larvae settling to bottom in control and adult-conditioned seawater

↓

Repeat 7 more times, using another 120 larvae per test

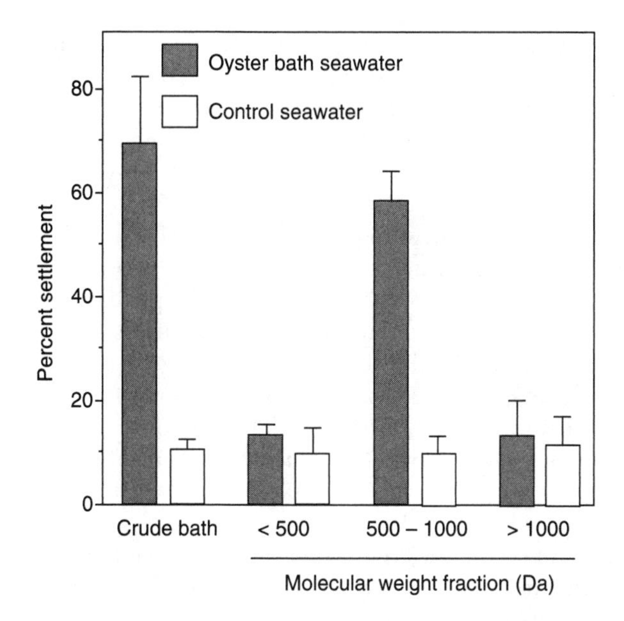

The active molecules have a molecular weight of 500-1000 daltons

Figure 35 The figure and take-home message for the first section of the poster shown in Figure 34.

The accompanying figure and associated take-home message for this first section are shown in Figure 35. **Each section of the poster will contain a separate Methods section and accompanying graph**, following the format just presented.

Making the Poster

Well in advance of the meeting, you will be told the dimensions of your display area—typically 4 feet high by 6 feet wide. Your entire display must fit within the designated area. **The title of your poster should be readable from 15 to 20 feet away**, so plan on using letters about 4 cm (~1.5 inches) tall. You can use a slightly smaller font for the names of all the authors and the institutions they are from. **The rest of your poster should be readable from about 4 to 6 feet away**, so text size should be about 1 cm (~3/8 inches) high.

If you print each component of the poster on a separate, 8.5-by-11-inch sheet of paper, then you can manipulate text size using your computer or an enlarging copying machine. Mount the individual items of your poster on colored paper or poster board. **Use a single background**

color for the entire poster to unify the presentation, perhaps using different shades to better distinguish the different sections of the poster. Choose a color that provides good contrast without being jarring or distracting; light brown and blue are good choices. Bring thumbtacks and tape to the meeting, in case they aren't provided at the site, so you can attach the components of your poster to the display board.

More commonly, posters are now prepared using PowerPoint, Adobe Illustrator, or similar software and then printed on a single, very large sheet of paper. Your school or a local business may have the facilities for printing such posters; you can also find good printing companies on the Internet and have the poster mailed to you. Again, bring thumbtacks to the meeting just in case.

When registering to present a poster or oral presentation at a meeting, you are generally required to submit an Abstract of your work (see p. 211), and **you may be required to include the Abstract in the upper left corner of your poster**. If so, be sure to leave room for the Abstract when planning your layout.

Creating a successful poster takes considerable planning, but it is well worth the time and effort required: Not only will you have a more interesting and enjoyable meeting, you will also return from the meeting with a permanent record of your study that can be displayed on a wall in the biology department.

CHECKLIST FOR MAKING POSTERS

- ❑ The poster includes all required information (e.g., institutional affiliations and names of all co-authors).
- ❑ All components will fit within the space provided for display.
- ❑ Title lettering is 4 cm tall, readable from a distance of at least 15 feet.
- ❑ Text letters are 1 cm high (~3/8 inches).
- ❑ The amount of text in each section is the minimum required.
- ❑ The flow of information on the poster is easy to follow.
- ❑ Methods are presented in flowchart form or as a simple listing.
- ❑ The Introduction states the specific issue that was addressed.
- ❑ Each figure or table is self-sufficient.
- ❑ Each major result is stated explicitly, as a take-home message.
- ❑ The poster has been checked for typographical and grammatical errors.
- ❑ If supplementary handouts are provided, they include the poster's title, names of all authors and their school affiliations, and the mailing and email addresses of the lead author.

12

WRITING LETTERS OF APPLICATION

An application for a job or for admission to a graduate or professional program will generally include a "CV" (for *curriculum vitae*) and an accompanying cover letter, both of which you write, as well as several letters of recommendation, which you generally never get to see. When applying to graduate or professional schools, and often when applying for jobs, you will also include a transcript of your college coursework and any special examination scores—for example, Graduate Record Examination (GRE) or Medical College Admission Test (MCAT) scores. You have no control over what your transcript and exam scores say about you; what is done is done. But you can still influence the message transmitted through your cover letter, résumé, and supporting letters. That influence works both ways: It can strengthen an otherwise weak case, or it can weaken an otherwise strong case.

Your CV summarizes your educational background, relevant work experience, relevant research experience (including a listing of publications if you have any), goals, and relevant skills and interests. The accompanying cover letter identifies the position for which you are applying and draws the reader's attention to the aspects of your CV that make you a particularly worthy candidate. The recommendations will give an honest assessment of your strengths and weaknesses (we all have some of each) and offer the reader an image of you as a person and as a potential employee or participant in a professional program. In this chapter, I will consider the art of preparing effective CV's and cover letters, and of increasing the odds of ending up with effective letters of recommendation.

BEFORE YOU START

Always try to put yourself in the position of the people who will be reading your application. What will they be looking for? They will probably be considering your application with 5 main questions in mind:

1. Is the applicant qualified for this particular position?
2. Does the applicant truly understand what the position entails?
3. Is the applicant really interested in our program or company?
4. Will the applicant fit in here?
5. What talents and abilities will this person bring to us?

Your application must address all 5 issues.

When you prepare your application, you should also consider that the number of applications received by a potential employer, professional school, or graduate program usually exceeds the number of positions available, often by a considerable margin. Many applicants will be qualified for the position, yet not every applicant can be interviewed or offered admission. Whoever begins reading your application will necessarily be looking for any excuse to disqualify you from the competition; your goal, then, must be to get the reader's interest at the start and hold it to the end.

PREPARING THE CV

The people who read your application will probably examine it for only a few minutes at most. Therefore, an effective CV is well organized, neat, and as brief as possible. It should not be longer than 2 pages.

There is no standard format for a CV; the model given in Figure 36 should be modified in any way that emphasizes your particular strengths and satisfies your own esthetic sense. However, **the CV is no place to be artsy or cute**; don't do anything that might suggest you are not taking the application process seriously. If you have an in-your-face email address like "ninjastudmuffin2@tufts.edu," change it before you apply for anything.

All CVs must contain the following 3 components:

1. Full name, address, telephone number, and e-mail address
2. Educational history
3. Relevant work, teaching, and research experience, if any

In addition, you will want to add any other information that makes you look talented, or well-rounded, or both:

4. Honors received
5. Papers published
6. Special skills
7. Outside activities, sports, hobbies

Robert Dick

Address and phone number

Until June 1, 2011:	After June 1:
P.O. Box 02933	Lakeview Drive
University Station	Narragansett, RI 02882
Castine, ME 04420	(401) 788-0153
(207) 201-1717	

Date of Birth: September 13, 1989

Goals: To earn a Ph.D. in Conservation Biology and pursue a career in teaching and research.

Education

Tufts University, Fall 2007-Spring 2011.
Major: Biology

Research Experience

Conducted a one-semester research project (in Dr. Oliver Hornbeam's laboratory) on the structure and function of guard cells in lyre-leaved sage, *Salvia lyrata* L., using transmission-electron microscopy. Presented the results of this research at the 10th New England Undergraduate Research Conference, University of New Hampshire, Durham, NH (April 2011).

Publications

Pechenik, J.A. and R. Dick. 2010. The influence of reduced pH on shell formation in the marine pteropod *Limacina helicina*. J. Wish. Think. 27:220-227.

Teaching Experience

Undergraduate teaching assistant for introductory biology laboratory. Fall semester 2009.

Honors

Elected to Phi Beta Kappa honor society, Spring 2010.
Received Churchill Prize in Biology (for performance in introductory Biology course). Fall 2007. Dean's List 7 out of 8 semesters.
Selected for teaching assistant position noted above.

Figure 36 Sample CV.

Work Experience

Summer 2004 Counselor, Lake Baker Summer Camp, AK
Summer 2005, 2008. Worked for Sweet Pea's Garden Center, Falmouth MA. Cared for all plants and shrubs, with one assistant.
Summer 2010. Assisted in the culture of oysters and hard shell clams. Mook Sea Farms, Damariscotta, ME.

Special Skills

Tissue preparation (fixation, embedding, sectioning) for transmission-electron microscopy.
Operation of JEOL Model 100CX transmission-electron microscope.
Developing 35-mm black and white film and TEM negatives; digital photography and image processing.

Outside Activities

Piano. URI Jazz band, 2008-2009
Swim team (2006-2010, captain 2010)
Campus tour guide Fall 2006, Fall 2010
Vegetable gardening (each summer since 2002)
Road racing (Boston Marathon 2008, 2010)

Figure 36 *(Continued)*

Avoid drawing attention to any potential weaknesses; if, for example, you lack teaching experience, do not write "Teaching experience: none." Use the CV exclusively to emphasize your strengths.

You might also add a 1- or 2-sentence statement of your immediate and long-range goals, if known, and the names of people who have agreed to write references on your behalf; this material is often incorporated into the cover letter instead, as will be discussed shortly.

You are not required to list age, race, marital status, height, weight, sex, or any other personal characteristic. **Be self-serving in deciding what to include.** If you think your youth might put you at a disadvantage, omit this information. If you think your age, race, or sex might give you a slight competitive edge, by all means include that information.

Do not be concerned if your first CV looks skimpy; it will fill out as the years go by. It is better to present a short, concise résumé than an obviously padded one.

You should **alter your CV for each application to focus on the different strengths required by different jobs or programs**. If, for example, Bob Dick, whose CV appears in Figure 36, were applying to a marine underwater research program, he might add under Special Skills that he is a certified SCUBA diver. If he were applying for a laboratory job in a hospital, he would probably omit the information about his diving certification.

PREPARING THE COVER LETTER

The cover letter plays a large role in the application process and is usually the first part of your application that an admissions committee or prospective employer reads. A well-crafted letter of application can do much to counteract a mediocre academic record. A poorly crafted letter, on the other hand, can do much to annihilate the good impression made by a strong academic performance. **Keep revising this letter** until you know it works well on your behalf. Have some friends, or perhaps an instructor, read and comment on your letter; then, revise it again. Be sure to computer-print the final copy; neatness counts, and a professional appearance also conveys seriousness of purpose. The time you put into polishing your cover letter is time well spent. The cover letter should be about 1 or, at most, 2 typed pages.

Do not simply write,

P.O. Box 66
University Center
Medford, MA 02155
May 1, 2012

Dear M. Pasteur:

I am applying for the position advertised in the *Boston Globe*. My résumé is enclosed. Thank you for your consideration.

Sincerely,

Earl N. Meyer

Earl N. Meyer

Although the letter ends well, its beginning is vague, and its midsection does nothing to further the applicant's cause. Use the cover letter to

1. Identify the specific position for which you are applying. (Monsieur Pasteur may have several positions open. Earl N. Meyer is applying for the position of research assistant, but how is M. Pasteur to know?)
2. Draw the reader's attention to the elements of your CV that you believe make you a particularly qualified candidate.
3. Indicate that you understand what the position entails and that you have the skills necessary to do a good job.
4. Convince readers that you are a mature, responsible person.
5. Convey a genuine sense of enthusiasm and motivation, but don't focus on what a wonderful opportunity the position will be for *you*. Many students assume that the faculty member or employer they are writing to is primarily interested in helping them, and in fact, most biologists do very much enjoy nurturing young investigators. However, we also have to think about how a potential new student can help *us* in our work—to what extent will this student contribute to the lab's productivity? In real life, active research requires outside funding (grant money), and grant money flows mostly to labs that ask good questions and get their results published. Promotions, too, and decisions about whether someone gets tenured or fired are often based at least partly on research productivity. So, instead of focusing on how much you can learn by working in someone's lab, **focus more on what you can bring to that lab that will keep it being productive.**

Before you begin to write the letter, ask yourself some difficult questions, and jot down some carefully considered answers:

Why do I want this particular job or to enter this particular graduate program?

What skills would be most useful in this job or program?

Which of these skills do I have?

What evidence of these skills can I present?

Your answers to these questions will provide the pattern and the yarn from which you will weave your cover letter. Not everyone will have a CV that looks like Bob Dick's (Fig. 36). But you need not have made the Dean's List every semester or have had formal teaching or research experience to impress someone with your application. In your cover letter, focus on the experiences that you *have* had. In lieu of teaching, perhaps you have done formal or informal tutoring. In lieu of formal research experience, perhaps you have taken numerous laboratory courses. Perhaps some experience you had in one or more of those laboratory courses influenced your decision to apply for a particular job or program. Perhaps acquiring certain skills in one or more of these laboratory courses has prepared you for the program or position for which you are applying. Or perhaps you can draw from experience outside biology to document your reliability, desire, and willingness to learn new things, or your ability to learn new techniques quickly. **We all have strengths; decide what yours are and, of those, which ones are appropriate for inclusion in your application.**

Tailor each letter to the particular position or program for which it is being prepared. **Try to find some special reason for applying to each program**; if possible, your application should reflect deliberate choice and a clear sense of purpose. If, for example, you have read papers written by a faculty member at the institution to which you are applying and have become interested in that person's research, weave this information into your cover letter. Should you take this approach, **you must say enough about that person's research or research area to convince readers that you understand what that person has done**. On the other hand, if your major reason for wanting a particular job or to attend a particular graduate program is the geographical location of the company or school, be careful not to state this as your sole reason for applying; as you write, and as you reread what you have written, try to put yourself on the receiving end of the cover letter, and consider how your statements might be interpreted.

Along these lines, if you are applying to a graduate program and hope to work with one or more particular professors, it's important to show how your research interests overlap with those of the targeted mentors. Graduate school is largely an apprenticeship program. There is no need to do exactly what the mentor or his or her students are currently doing, but there should be some overlap of interests: During your studies, you will need access to relevant equipment, feedback, and advice, and to get those things you'll want to work with people in your general field of interest. Before writing your cover letter, and before contacting any potential mentor, be sure to look carefully at that person's Web site and read about his or her current research activities, and look carefully at some of his or her research publications to be sure that the fit will be a good one for you both.

Back up all statements with supporting details. Avoid simply saying that you have considerable research experience. Instead, briefly explain what your research experience has been. Do not state that you are a gifted teacher; describe your teaching experience. Similarly, don't just say that you're creative in the laboratory or hardworking; give an example of that creativity or motivation and let readers come to the right conclusion from the facts that you present.

Sign the letter with your given name, not a nickname; again, don't run the risk of not being taken seriously.

Here is an example of a weak cover letter. Similar letters have, unfortunately, been submitted by people with very good grades, test scores, and letters of recommendation. The author is applying for admission to a Ph.D. program in Biology.

P.O. Box 666
University Center
Medford, MA 02155
May 1, 2012

To the admissions committee:

I have always been fascinated by the living world around me. I marvel at the details of the way biology works, and I would now like to fulfill my curiosity and passion for biology in pursuing a Ph.D. in your program.

As you can see from my transcript, I have taken 12 courses in biology (2 more than the number needed for graduation) and have done well in most of them. I am very interested in the work that Professor Orians does with terrestrial plants and would especially enjoy studying the physiological effects of humans and pollution. In my senior year, I did a small project examining the effects of increased carbon dioxide levels on local plants.

I would also like to apply for the teaching assistantship award. I have always liked helping people learn about science, and I am eager to communicate my enthusiasm for biology to others. I have requested that my GRE scores be sent directly to you.

I look forward to your reply.

Sincerely,

Variola Major

Variola Major

Remember, the admissions committee is looking for any excuse to disqualify applicants; this letter gives the committee just that excuse, regardless of what the rest of the application looks like. **The letter conveys enthusiasm, but a very naive enthusiasm.** What does the student find interesting about the physiology of plants? What was the research project? What question was asked? How was the question addressed? What results were obtained? Did the student learn anything from the experience? Does the student really know anything about plant physiology? Does she have any detailed understanding of what Professor Orians actually does? Has she read any of his papers? What makes the student think she would be an effective teacher? Has she had any teaching experience? Does she understand what teaching entails?

The rest of the application letter is equally uninformative. Why is the student applying to this particular program? All biology majors take biology courses, and the student's grades are already on the transcript. What has the student learned from these courses that makes her want to pursue advanced study?

The same student could have written a much more effective letter by thinking about what the admissions committee might be looking for and by documenting her strengths. Here is an example of the way this student might have rewritten her letter:*

*The research described in her letter is based on a paper by Orians, C.M., Floyd, T. 1997. *Oecologia*, 109; 407–413.

P.O. Box 666
University Center
Medford, MA 02155
May 1, 2012

To the admissions committee:

Please consider my application for admission to your Ph.D. program in Biology. I will be graduating from Mordor University in May with a B.S. in Botany. I believe that I have the experience and motivation to make a contribution to your program.

I became interested in plant physiological ecology through a seminar course taught by Professor Mendel. This was my first experience reading the original scientific literature, but by the end of the semester, I was able to present a well-received research proposal on the subject of root growth, based largely upon my own library research.

Dr. Mendel later invited me to participate in a research project examining the relationship between plant nutrition and resistance to attack by 4 insect species. I conducted this research in the field using 2 common willow species (*Salix eriocephala* and *S. sericea*). The goal was to determine whether the well-documented variation in susceptibility to attack might be explained by differences in soil nutritional quality, which we manipulated by applying different amounts of fertilizer.

To begin, we distributed cuttings from each species among 3 treatments differing in the concentration of fertilizer provided each week, with 9 replicates per treatment. At intervals over the next 4 months, I counted the number of leaves and shoots on each plant and measured their average sizes to document the effects of the nutrient treatments on plant growth rates. On 2 occasions we also quantified the degree of damage caused to the plants by 2 species of leaf-mining caterpillar, a leaf-folding sawfly, a leaf-chewing beetle, and a fungal pathogen.

It turned out that for both willow species, the more fertilizer the plants received, the faster they grew and the more susceptible they were to attack by most of the pests. I also discovered, somewhat surprisingly, that the 2 plant species differed dramatically in their susceptibility to the different pests. Our working hypothesis is that the 2 willow species probably differ in the amounts of phenolic glycosides and other defensive compounds contained in their leaves, and that within a species, these compounds are produced at lower concentrations in faster-growing plants.

Through this study, I learned the importance of careful experimental design and data analysis, and, more important, that I have the patience to do research. I do not wish to commit myself to a specific field of research at this time, believing that I would benefit from an additional year of literature and laboratory exploration, but I believe that I would like to explore the potential impact of human activities on the ability of plants to produce chemical defenses. During my first year, I would hope to take your 2 upper-level courses in plant physiology and biochemistry and an additional statistics course (in analysis of covariance).

During my last semester at Mordor University, I have been serving as an undergraduate teaching assistant for the introductory biology laboratory. I find that having to explain things to other students forces me to come to grips with what I do and do not know. I am enjoying the challenge greatly, and I look forward to doing additional teaching in the future; I am learning a lot about biology through teaching.

I have asked the following faculty members for letters of recommendation:

Professor G. Mendel (Biology Dept., Mordor University)
Professor Jonathan Allen (Randolf Macon College)
Professor Todd Nickle (Mount Royal College)

Thank you for considering my application. I look forward to receiving your response.

Sincerely,

Variola Major

Variola Major

This letter, too, conveys enthusiasm, but it is an enthusiasm that reflects knowledge, experience, maturity, and commitment. The applicant seems to understand what research is all about and apparently knows how to go about doing it. Moreover, we see that Ms. Major thinks clearly and writes well.

Variola would write a somewhat different letter if she were applying for a job as a technician in a research laboratory. In this letter, she would want to emphasize her skills and reliability as a laboratory worker and her interest in the type of research being done in the laboratory to which she is applying. An example of such a letter follows:

Both of these letters convey knowledge of the position or program applied for, a sincere interest in biology, and a high level of ability and commitment. Your letter should do the same. Your credentials may not be as impressive as Ms. Major's, but if you think about the experiences you have had in relationship to the skills required for the position or program, you should be able to construct an effective letter. Take your letter through several drafts until you get it right (see Chapter 6, on revising).

P.O. Box 666
University Center
Medford, MA 02155
May 1, 2012

Dear Professor Hornbeam:

Please consider my application for the technical position you advertised recently in the *Boston Globe*. I will be graduating from Mordor University this May with a B.S. degree in botany. Although I eventually expect to return to school to pursue a Ph.D., I would first like to work in a plant physiological ecology laboratory for 1 or 2 years to learn some additional techniques and to become more familiar with various research fields and approaches.

I first became interested in plant physiology through a seminar course with Professor G. Mendel at Mordor University. During the semester, we read 2 of your recent papers describing the effects of hybridization on leaf biochemistry in goldenrod.

Following this seminar, I began studying (under the direction of Dr. Mendel) the influence of soil nutrient quality on the susceptibility of willow plants *(Salix eriocephala* and *S. sericea)* to damage by leaf-mining caterpillars and fungi. We found that the more fertilizer the plants received, the faster they grew and the more susceptible they were to attack. The next step in the study will be to determine whether the differences in pest resistance can be explained by other chemical defenses by the plant's leaves.

Through this study, I learned a variety of general laboratory techniques (use of assorted balances, sterile culture methodology, how to digitize images for quantitative analysis of leaf size, and how to use 2 major statistical programs—GraphPad Prism and Systat). I also learned that I have the patience and motivation needed to do careful research. I have taken 6 laboratory courses in biology (General Genetics, Invertebrate Zoology, Comparative Animal Physiology, Plant Physiology, Cell Biology, and Developmental Biology), in which I learned several specialized laboratory techniques, including the pouring and use of electrophoretic gels, measurement of organismal and mitochondrial respiration rates, and use of a vapor pressure osmometer. I also learned to design quantitative field studies in terrestrial ecosystems.

In short, I am very much interested in your research and believe I can make a contribution to the work of your laboratory. I have requested letters of recommendation from the following faculty at Mordor University:

Professor G. Mendel (Biology Dept.)
Professor Phillip Starks (Biology Dept.)
Professor Juliet Fuhrman (Biology Dept.)

Thank you for considering my application; I look forward to hearing from you.

Sincerely,

Variola Major

Variola Major

EFFECTIVE LETTERS OF RECOMMENDATION

Letters of recommendation can be extremely important in determining the fate of your application. Although you do not write these letters yourself, and rarely even get the opportunity to read them, you can take steps to increase their effectiveness.

Getting an A in a course does not guarantee a helpful letter of recommendation from the instructor of that course. The most useful letters to admissions committees and prospective employers are those commenting on the following characteristics: laboratory skills, communication skills (written and oral), motivation, ability to use time efficiently, curiosity, maturity, intelligence, ability to work independently, and ability to work with others. Instructors cannot comment on these attributes unless you become more than a grade in their record books. Make an appointment to talk with some of your instructors about your interests and plans. We faculty members are usually happy for the opportunity to get to know our students better.

When it is time to request letters of recommendation, choose 3 or 4 instructors who know something about your relevant abilities and goals, and ask each of them if he or she would be able to support your application by writing a letter of recommendation. Give each person the opportunity to decline your invitation. If the people you ask agree to write on your behalf, make their task easier by giving them a copy of your CV; transcript; letter of application; and, if appropriate, the job advertisement itself, and be sure that your champions understand why you want to do what you are proposing to do. **Be certain to indicate clearly the application deadline** and the mailing address or e-mail address to which the recommendation should be sent.

It takes as much time and thought to write an effective letter of recommendation as it takes to write an effective letter of application. Don't lose goodwill by requesting letters at the last minute. **Give your instructors at least 2 weeks** to work on these letters. "It has to be in by this Friday" will probably annoy your prospective advocate and, even if he or she is still in a cooperative mood, may not allow the recommender the time needed to prepare a good letter. Moreover, last-minute requests don't speak favorably about your planning and organizing abilities, and they imply a lack of respect for your instructor. So, be considerate and thereby get the best recommendation possible.

Appendix A

Revised Sample Sentences in Final Form

1. ~~To perform this experiment there had to be a low tide.~~ We conducted the study at Blissful Beach on September 23, 1991. at ~~2:30 PM.~~ at low tide

   We conducted the study at Blissful Beach at low tide on September 23, 1991.

2. In *Chlamydomonas reinhardi*, a single-celled green alga, there are two matine types, + and -. The + and - cells mate with each other when starved of nitrogen and form a zygote. g

   In *Chlamydomonas reinhardi*, a single-celled green alga, there are two mating types, + and −. When starved of nitrogen, the + and − cells mate with each other and form a zygote.

3. Protruding form this carapace is the head, bearing a large pair of second antennae.

   Protruding from this carapace is the head, bearing a pair of large second antennae.

4. The order in which we think of things to write down is rarely the order we use when ~~we~~ explain what we did to a reader. ing

   The order in which we think of things to write down is rarely the order we use when explaining to a reader what we did.

5. ~~The purpose of~~ Professor Wilson's book ~~is the~~ examin~~ation~~ ~~of~~ questions of evolutionary significance. es

   Professor Wilson's book examines questions of evolutionary significance.

6. Swimming ~~in fish~~ has been carefully studied in only a few species.

   The mechanics of swimming have been carefully studied for only a few fish species.

7. One example of this capcity is ~~observed in~~ the ~~phenomenon of~~ encystment exhibited by many fresh water and parasitic species.

   One example of this capacity is the encystment exhibited by many freshwater and parasitic species.

8. In a sense, then, the typical protozoan ~~may be regarded as being~~ a single-celled organism.

   In a sense, then, the typical protozoan is a single-celled organism.

9. An estuary is a body of water nearly surrounded by land whose salinity is influenced by freshwater drainage.

   An estuary is a body of water nearly surrounded by land and whose salinity is influenced by freshwater drainage.

10. The carbon-to-nitrogen ratio of microbial films ~~gives an indication of the film's~~ nutritional quality (A. Bhosle and C. Wagh, 1997)

    The carbon-to-nitrogen ratio of microbial films indicates their nutritional quality (Bhosle and Wagh, 1997).

11. In textbooks and many lectures, ~~you are being~~ presented with facts and interpretations.

    Textbooks and many lectures present you with facts and interpretations.

12. The human genome contains about 25,000 genes, however there is enough DNA in the genome to form nearly 2 x 10^6 genes

    The human genome contains about 25,000 genes; however, there is enough DNA in the genome to form nearly 2×10^6 genes.

13. ~~It should be noted that~~ analys~~es~~ ~~were done~~ to determine whether the caterpillars chose the different diets at random.

    The data were analyzed to determine whether the caterpillars chose the different diets at random.

14. These experiments ~~were conducted to~~ test whether the ~~condition of the~~ biological films ~~on the substratum surface~~ triggered settlement ~~of the larvae.~~

    These experiments tested whether the biological surface films triggered larval settlement.

15. ~~Various species of sea anemones live throughout the world.~~

    (Sentence deleted for lack of content.)

16. Th~~is~~ data clearly demonstrates that growth rates of the blue mussel (mytilus Edulis) vary with temperature.

    These data clearly demonstrate that growth rates of the blue mussel (*Mytilus edulis*) vary with temperature.

17. Hibernating mammals mate early in the spring ~~so that~~ their offspring ~~can~~ reach adulthood before the beginning of the next winter.

    Hibernating mammals mate early in the spring. As a consequence, their offspring reach adulthood before the beginning of the next winter.

18. This study ~~pertains to the investigation of~~ the effect of this pesticide on the orientation behavoir of honey bees.

    This study describes the effect of this pesticide on the orientation behavior of honey bees.

19. ~~The results reported here have lead the author to the conclusion that~~ thirsty flies ~~will~~ show a positive response to all solutions, regardless of sugar concentration (~~see~~ figure 2).

    Thirsty flies apparently show a positive response to all solutions, regardless of sugar concentration (Fig. 2).

20. Numbers are difficult for listeners to keep track of ~~when they are~~ floating around in the air.

    Numbers floating around in the air are difficult for listeners to keep track of.

21. Those seedlings ~~possessing a quickly growing phenotype~~ genetically programmed for faster growth will be selected for, whereas. . .

    Those seedlings genetically programmed for faster growth will be selected for, whereas …

22. ~~Under~~ a dissecting microscope, a slide with a drop of the culture was examined at 50x. using

    A slide with a drop of the culture was examined at 50× using a dissecting microscope.

23. Measurements of salamander respiration ~~by the salamanders~~ typically took one-half hour each.

    Measurements of salamander respiration typically took one-half hour each.

24. ~~The~~ results suggest that ~~some local enhancement of~~ pathogen-specific antibody ies are produc~~tion~~ ed at the infection site ~~exists~~. and thus are enhanced locally.

    Results suggest that pathogen-specific antibodies are produced at the infection site and thus are enhanced locally.

25. ~~Usually it has been found that higher temperatures (30°C) have resulted in the production of females, while lower temperatures (22–27°C) have resulted in the production of males. (e.g., Bull, 1980; Mrosousky. 1982)~~

    The turtles are typically born female when embryos are incubated at 30°C, and male when incubated at lower temperatures (22-27°C) (e.g., Bull, 1980; Mrosovsky, 1982).

    The turtles are typically born female when embryos are incubated at 30°C, and male when incubated at lower temperatures (22–27°C) (e.g., Bull, 1980; Mrosousky, 1982).

26. Octopuses have been successfully trained to distinguish between red and white balls of ~~varying~~ different size s.

    Octopuses have been successfully trained to distinguish between red and white balls of different sizes.

Appendix B

Commonly Used Abbreviations

	ABBREVIATION	EXAMPLE
Length		
meter	m	3 m
centimeter (10^{-2} meter)	cm	15 cm
millimeter (10^{-3} meter)	mm	4.5 mm
micrometer (10^{-6} meter)	μm	5 μm
Weight		
gram	g	10 g
kilogram (10^3 grams)	kg	15 kg
milligram (10^{-3} gram)	mg	16 mg
microgram (10^{-6} gram)	μg	4 μg
nanogram (10^{-9} gram)	ng	8 ng
picogram (10^{-12} gram)	pg	11 pg
femtogram (10^{-15} gram)	fg	10 fg
Volume		
liter	l or L	3 l or 3 L
milliliter (10^{-3} liter)	ml or mL	37 ml or 37 mL
microliter (10^{-6} liter)	μl or μL	13 μl or 13 μL
Time		
months	mo	6 mo per year
weeks	wk	4 wk
days	d	2 d
hours	h	Wake me up in 24 h
minutes	min	20 min
seconds	s	60 s
Concentration		
milliosmoles/liter	mOsm L^{-1}	650 mOsm L^{-1}
mole	mol	0.13 g mol^{-1}
molar	mol L^{-1}	a 0.3 molar solution
salinity (parts per thousand)	‰ S, ppt	31‰ S seawater or 31 ppt
practical salinity units	psu	28.5 psu
parts per million	ppm	0.2 ppm copper
parts per billion	ppb	200 ppb copper
Statistics		
mean	$\overline{X}$	$\overline{X} = 27.2$ g $individual^{-1}$
standard deviation	SD	SD = 0.8
standard error	SE	SE = 0.3
sample size	N	N = 16
p-value	*p*	$p < 0.01$
Other		
newton (a measure of force)	N	26.1 N
joule (a measure of work or energy)	j	25.5 j (= 25.5 N·m)
photoperiod (h light: h dark)	L:D	10L:14D
1 species	sp.	*Crepidula* sp.
2 or more species	spp.	*Crepidula* spp.
approximately	c., ≈	c. 25°C, or ≈25°C

Appendix C

Recommended Resources

BOOKS AND ARTICLES ABOUT SCIENTIFIC WRITING

Davis, M. 2005. *Scientific papers and presentations*, 2nd edition. New York: Academic Press.

Day, R.A. 1992. *Scientific English: A guide for scientists and other professionals*. Phoenix, AZ: Oryx Press.

Day, R.A. 1998. *How to write and publish a scientific paper*, 5th edition. Phoenix, AZ: Oryx Press.

Gopen, G.D., Swan, J.A. 1990. The science of scientific writing. *American Scientist, 78*, 550–558.

King, L.S. 1978. *Why not say it clearly? A guide to scientific writing*. Boston, MA: Little, Brown.

Knisely, K. 2009. *A student handbook for writing in biology*, 3rd edition. Sunderland, MA: Sinauer Associates and W.H. Freeman.

O'Connor, M. 1991. *Writing successfully in science*. New York: Chapman & amp; Hall.

Penrose, A.M., Katz, S.B. 2003. *Writing in the sciences: Exploring conventions of scientific discourse*. New York: St. Martin's Press.

Wilkinson, A.M. 1991. *The scientist's handbook for writing papers and dissertations*. Englewood Cliffs, NJ: Prentice Hall.

Zinsser, W. 1995. *On writing well: An informal guide to writing nonfiction*, 4th edition. New York: HarperCollins.

TECHNICAL GUIDES FOR BIOLOGY WRITERS

Council of Science Editors. 2006. *Scientific style and format: The CSE manual for authors, editors, and publishers*, 7th edition. New York: Cambridge University Press.

ADVICE ON ANALYZING DATA AND CONSTRUCTING EFFECTIVE GRAPHS

Cleveland, W.S. 1994. *The elements of graphing data*, 2nd edition. Summit, NJ: Hobart Press.

Motulsky, M. 1995. *Intuitive biostatistics*. New York: Oxford University Press.

Parkhurst, D.F. 2001. Statistical significance tests: Equivalence and reverse tests should reduce misinterpretation. *BioScience 51*, 1051–1057.

Quinn, G.P., Keough, M.J. 2001. *Experimental design and data analysis for biologists* (Chapter 19, pp. 494–510, on graphing data). New York: Cambridge University Press.

Tufte, E.R. 2001. *The visual display of quantitative information*, 2nd edition. Cheshire, CT: Graphics Press.

WEB SITES

Advice on Grammar and Punctuation

http://owl.english.purdue.edu/handouts/
(Select "Grammar, punctuation, spelling")

http://writing-program.uchicago.edu/resources/grammar.htm

http://grammar.ccc.commnet.edu/grammar/marks/quotation.htm
(Fantastic! Everything you could possibly want to know about the use of quotation marks!)

Locating Useful References

http://owl.english.purdue.edu/owl/resource/558/01/
(Searching the World Wide Web: Overview_)

http://ase.tufts.edu/biology/bguide/
(A Biologist's Guide to Library Resources)

http://www.sc.edu/beaufort/library/pages/bones/bones.shtml
(Barebones 101: A Basic Tutorial On Searching the Web)

http://www.lib.berkeley.edu/TeachingLib/Guides/Internet/FindInfo.html
(How to Choose the Search Tools You Need, from UC Berkeley)

http:www.ipl.org/
(A merging of resources from the Internet Public Library and the Librarian's Index to the Internet)

http://sib.illinois.edu/SkillGuidelines/

Evaluating Web Sites

http://lib.nmsu.edu/instruction/evalcrit.html
http://www.vuw.ac.nz/~agsmith/evaln/index.htm
http://www.ithaca.edu/library/training/think.html
http://www.library.tufts.edu/tisch/webeval.htm
http://www.lib.berkeley.edu/TeachingLib/Guides/Internet/Evaluate.html

Citing Web Sources

http://www.nlm.nih.gov/pubs/formats/internet.pdf
http://www.lib.berkeley.edu/instruct/guides/citations.html
http://ia.juniata.edu/citation/cse/

Getting the most from Excel

http://sib.illinois.edu/SkillGuidelines/ExcelGuidelines2003.html
(Using Excel to make figures and to do simple statistical tests)

Understanding and Using Statistics

http://www.graphpad.com/
(Click on Resource Library)
http://sib.illinois.edu/SkillGuidelines/BasicStatistics.html

Writing Research Papers

http://www.library.tufts.edu/researchpaper
http://www.ruf.rice.edu/~bioslabs/tools/report/reportform.html
http://abacus.bates.edu/~ganderso/biology/resources/writing/HTWgeneral.html
http://sib.illinois.edu/SkillGuidelines (choose Writing a Scientific Manuscript, or Sample Manuscript with commentary)

Giving Effective Powerpoint and Other Oral Presentations

http://colinpurrington.com/tips/academic/sciencetalks
(Courtesy of Professor Colin Purrington)
http://www.kumc.edu/SAH/OTEd/jradel/effective.html
(Courtesy of Professor Jeff Radel, Univ. Kansas Medical Center)

Preparing Effective Poster Presentations

http://colinpurrington.com/tips/academic/posterdesign
(Courtesy of Professor Colin Purrington)
http://www.kumc.edu/SAH/OTEd/jradel/effective.html
(Courtesy of Professor Jeff Radel, Univ. Kansas Medical Center)
http://library.buffalo.edu/asl/guides/bio/posters.html

Organizing Your Time

http://www.library.tufts.edu/researchpaper
(or Google "research paper navigator Tufts")

Index